The Laws of Robots

Law, Governance and Technology Series

VOLUME 10

For further volumes:
http://www.springer.com/series/8808

Ugo Pagallo

The Laws of Robots

Crimes, Contracts, and Torts

Springer

Ugo Pagallo
University of Torino
Torino Law School
Torino, Italy

ISBN 978-94-007-6563-4 ISBN 978-94-007-6564-1 (eBook)
DOI 10.1007/978-94-007-6564-1
Springer Dordrecht Heidelberg New York London

Library of Congress Control Number: 2013937543

Printed on acid-free paper

Springer is part of Springer Science+Business Media (www.springer.com)

To Alexis, Anna Sofia, and the Next Generation

Preface

I am inside the orbit of Deimos and completely on my own. Wish me luck!

Curiosity Mars, tweeting on 5 August 2012 at 8:12 p.m., US Pacific Time (two hours and twenty minutes before the robotic rover successfully landed on the red planet).

The year 1961 was notable and moreover, a turning point for one of the most breath-taking fields of today's information revolution, robotics. The amazing pace in the field of robotics and its manifold applications can be traced back to 1961 and the remarkable sequence of events concerning politics, military confrontations, scientific research, culture, society, and the progress of technology. To put things in context, on 12 April 1961, Yuri Gagarin became the first man in space, soon followed by US Navy Commander Alan Shepard on 5 May. In between, about 1300 Cuban exiles armed with US weapons, and sponsored by the CIA, landed at the Bay of Pigs on 17 April, unsuccessfully attempting to overthrow Fidel Castro's regime. Four months later, on 17 August, East Germany (DDR) started to erect the Berlin Wall. A few weeks later, at 11:32 a.m. on 30 October, the USSR detonated the Tsar Bomb, causing the largest man-made explosion in history, namely a 50-megaton hydrogen bomb over the Novaya Zemlya archipelago. Luckily, in this hottest of years during the cold war, technology and science also advanced for more peaceful purposes: Squibb produced the first electric toothbrush, movies were shown for the first time on TWA flights, IBM presented its Selectric typewriter, and Jack Lippes developed the contraceptive intrauterine device. While some glorious movies, such as *West Side Story*, *Breakfast at Tiffany's* and *La Dolce Vita*, were released, a number of unforgettable songs like "Stand by Me," or "Hit the Road Jack," made the charts.

In addition to the publication of such famous books as *Tropic of Cancer* and *The Winter of Our Discontent*, 1961 is also the year when some famous baby boomers were born, such as President Barrack Obama, the jurist Larry Lessig, Princess Diana, George Clooney, Eddie Murphy and, yes, the Fantastic Four: Mister Fantastic, the Invisible Woman, the Human Torch and the monstrous Thing. For that matter, the author of this book was also born in 1961, just in time to enjoy the first disposable diapers, *i.e.*, Pampers.

Besides FM stereos, the new Coca Cola rival of 7 Up, Sprite, and Johnson & Johnson's Tylenol, we should not miss another novelty brought on in 1961. Forty-one years after the word "robot" became popular with Karel Ĉapek's play *Rossum's Universal Robots* (1920), and almost 20 years after Isaac Asimov coined the term "robotics" in his novel *Runaround* (1942), robots were employed in the industry sector for the first time. Contrarily to Ĉapek's humanoids and Asimov's artificial agents, these machines were neither robot soldiers nor spacewalkers. Rather, the first industry robot was tested within the automobile sector, drawing on the projects of George Devol and Joseph Engelberger, which culminated in the UNIMATE robot performing spot welding and extracting die-castings in a General Motors factory in New Jersey. Soon after, the idea was not only to manufacture machines (*e.g.*, cars) through further machines (*e.g.*, robots). The plan was to build fully autonomous cars, later dubbed as unmanned ground vehicles, or "UGVs," according to several different projects pursued in the USA, Japan, Germany and Italy.

Yet, it was only 20 years later, in the early 1980s, that the use of robotics within the car industry became critical. Japanese industry first began to implement this technology on a large scale in their factories, acquiring strategic competitiveness by decreasing costs and increasing the quality of their products. As this was the time of my first lengthy stay in Silicon Valley, I vividly recall the sense of shock instilled by the first wave of Japanese automobiles overwhelming Detroit cars on the Californian speedways in summer 1982. Western car producers learned a hard lesson and followed Japanese thinking, installing robots in their factories a few years later. This massive trend went on for two decades: remarkably, in the *Editorial* to the World 2005 Robotics Report of the Economic Commission for Europe and the International Federation of Robotics, Åke Madesäter raised the risk that the robot industry was too focused and dependent on the automotive industry: "The industrial robot industry has become too dominated by car manufacturers and its sub-suppliers. In the period 1997–2003, the automotive industry in Spain received 70 % of all new robot installations. In France, the United Kingdom and Germany the corresponding figure amounted to 68 %, 64 % and 57 %, respectively" (UN 2005: ix).

In the same years as covered by the UN World report, however, things began to rapidly change: the two decade dependence of robotics on the automobile industry dramatically opened up to diversification, a revolution as phrased by scholars. This occurred with water-surface and underwater unmanned vehicles, or "UUVs," used for remote exploration work and the repairs of pipelines, oil rigs and so on, developing at an amazing pace since the mid-1990s. A decade later, unmanned aerial vehicles ("UAVs"), or systems ("UAS"), upset the military field. As the *U.S. Army Unmanned Aircraft Systems Roadmap 2010–2035* illustrates, their quantitative and qualitative indices are impressive. From 2003 to 2008, UAV flights increased by 2,300 % and the number of UAVs, which was less than 50 before 2001, was over 3,000 in 2006, over 7,000 in 2010, and well over 12,000 at the time of this writing. The impact of UAVs on the laws of war has given rise to UN special rapporteurs and scholars alike proposing stricter regulations for their use. Whereas "the difference between science fiction and science is timing," in the phrasing of the Colonel Christopher B. Carlile, Director of the UAS Center of Excellence in Fort Rucker, Alabama, it is no surprise then that the Sci-Fi menace of Ĉapek's robot soldiers in *R.U.R.* has turned out to be real.

After the UUV and UAV revolutions with their normative challenges, *e.g.*, swarms of tiny drones that plan the missions they are going to execute by themselves, further candidates for the next robotic revolution are a new generation of UGVs, that is, smart cars driving themselves on the highways in fully autonomous, or semi-autonomous, ways. A number of states, organizations and private companies have seriously pursued this project over the past years. Contemplate the Grand Challenge competitions organized by the US Defence Advanced Research Projects Agency ("DARPA") since the late 1990s. Among the participants of such challenges, suffice it to mention the vehicles sponsored by Carnegie Mellon with General Motors, Stanford with Volkswagen, and Google's driverless cars. After the Eureka Prometheus Project (1987–1995), the European Commission has similarly promoted the "Intelligent Car Initiative" in 2010, in order to drastically reduce traffic jams and car accidents, while improving energy efficiency and polluting less. Certain terrifying figures can make us fully appreciate that which is at stake with the next UGV revolution: road transport accounts for more than one-quarter of the EU's total energy consumption, costs of traffic jams amount to approximately 0.5 % of EU GDP, car congestions impact 10 % of the European major road networks, in which there are around 1.3 million mishaps and 41,000 people who die in car accidents every year.

The panoply of robotic applications available suggests further candidates for the next robotic revolution. Reflect on the set of applications for personal and domestic service: we already have a number of robot toys and robot nannies that are programmed to provide love and take care of children and the elderly. In the academic field, think of a new generation of artificial assistants for university teachers, as a sort of i-Jeeves that could help us schedule conferences, lectures and meetings. By checking the availability and convenience of logistics in accordance with a number of parameters like budget, time efficiency, or weather average conditions, the robot could report its findings back for a decision or, even, determine the steps of the academic tour by directly accepting invitations, booking hotel rooms, flights and so forth. Moreover, we should take into account the class of robotic scientists that may independently discover new knowledge without the need for human intervention, as occurred with "Adam" at Aberystwyth University and the University of Cambridge in 2009, when researchers confirmed that such a robot discovered new evidence about the genomics of the baker's yeast *Saccharomyces cerevisiae*. Likewise, consider NASA's mars rover robots and the Science Laboratory flight team: the one-ton, $ 2.5 billion machine became especially popular on 5 August 2012 when the robot, Curiosity, using a supersonic parachute and a first-of-its-kind "sky crane," successfully landed on the red planet to discover more about the martial environment and reach places scientists deem as interesting for further study.

Another amazing class of robotic applications concerns hybrids of natural and artificial systems, much as machines that mimic animals and their behaviour. Although nature has required billions of years to refine its own design, so that many of the ideas on animal-like robotic behaviour often outpace the capacities of today's technology, several interesting projects are on their way: robots that exploit the design choices of multi-objective ant colonies or of brood comb constructions by the stingless bees, up to the development of unmanned micro-drones that fly like an albatross. Whilst hybrids of natural and artificial systems include such applications as nanorobots controlled by muscle cells, or neuroprostheses translating the thought of quadriplegics, the troubles with the computing power of robots have increasingly been addressed by connecting them to a networked repository on the internet, allowing robots to share the information required for object recognition, navigation and task completion in the real world. As part of the *Cognitive Systems and Robotic Initiative* from the European Union seventh framework programme (FP7/2007–2013), this is the aim of the RoboEarth project as a world wide web for robots, that is, a network and database repository where machines can share information and learn from each other about their behaviour and their environment. Avoiding the shortcomings of traditional approaches, such as on-board computers for robots, the goal is to complete a sort of cloud robotics infrastructure with all that is needed to close the loop from robots to RoboEarth to robots.

In addition to further examples, *e.g.*, AI soccer players, what this panoply of robotic applications makes clear is a paramount aspect of the current information revolution, namely the astonishing exponential pace of innovation and technological progress after two decades of a too dependent car industry sector. This acceleration is usually illustrated, or even summed up, with the "Moore's law," *i.e.*, the 1965 self-fulfilling prophecy that the computing power of chips would have doubled every 18 months. In addition to the economical, political, and cultural conditions that may favour the use of a certain technology, the almost five decades-long rates of doubling amounts of computation have not only made feasible what simply was impossible few years before, but have opened up new horizons of further technological development. To clarify this point, let me recall a family story that involves one of the most spectacular flops of Apple's history, that is, the 1992 personal digital assistant Newton. This sort of proto i-Pad with touch-screen and pen-stylus included some applications, as "names," "dates," and "notes," much as simple tools as time zone maps, currency converter, and calculator, that allowed users to gather, manage, and share their information. Contrary to the i-Pad, however, the reason why, at least for my sister and her colleagues, Newton turned out to be a failure mostly depended on the fact that such Apple devices simply arrived 15 years too early and, frankly, were too expensive. Returning to the field of robotics, and by further considering a number of factors such as public research and development (R&D) support, interagency transfers, and growing access to powerful and cheaper software and hardware, we can thus understand a simple truth: whereas each of the initial leaps in the realm of robotics required a 20-year interval, nowadays it seems that almost every year brings about some sort of robotic revolution. From Asimov's *Runaround* to the current Mars rover machines, a 70-year old story of robotics can be summarized as a classic symphony in four movements.

First, *adagio ma non troppo*: industrial robots were introduced in the manufacturing sector in 1961, that is, almost 20 years after Asimov's first novel on robotics. Second, *andante con brio*: the use of robots within the car industry became critical in the early 1980s, that is, 20 years after the introduction of the first industrial robot in the automobile field. Third, *ostinato*: in the early 2000s, certain individuals still had the impression that robotics was too dependent on the automobile industry. Fourth, much as at the end of Beethoven's ninth symphony, *prestissimo, maestoso, molto prestissimo*: both the quantity and quality of robotics applications have somehow spiralled out of control in the past decade, so much so that the exponential curve of advancement in the field of robotics has given rise to certain exaggerations. In light of the new generation of driverless cars, UAVs and UUVs, robotic scientists, hybrids of natural and artificial systems, and so forth, advocates of the techno-deterministic stance argue that the current information revolution inexorably shapes the destiny of human beings and their societies, so that intelligent machines will succeed humans and we, as a species, could face extinction. Greater than human intelligence, in other words, will emerge through nanobots, artificial intelligence and robotics, as the main contributing factors to this singular event.

However, we do not have to perceive the advancement of robotics as inexorable as the revolutionary movement of the planets, to acknowledge that a number of robotic applications transform and reshape individual and social environments through a new set of constraints and opportunities. The panoply of such robotic applications entails nevertheless a high degree of specialization, suggesting that we should avoid any sort of broad-brush stroke illustration of the topic. Robotics traditionally draws on such disciplines as engineering and cybernetics, artificial intelligence and computer science, physics and electronics, biology and neuroscience, down to the fields of humanities: politics, ethics, economics, law, etc. The extraordinary variety of robotic applications, on one hand, cautions us against generalizations that would inescapably fall short in determining, say, the normative challenges of the field. Whereas, for example, it is likely that drones and other types of autonomous (lethal) weapons mainly affect such fields as international humanitarian and criminal law, other applications, such as da Vinci robot-surgeons, mostly raise matters of contractual obligations and strict liability rules.

On the other hand, the multi-disciplinary nature of robotics suggests that an all-encompassing view of the field far exceeds the capacities of a single scholar. When Massimo Durante and myself were planning a book on legal informatics and the normative challenges of technology in 2011, we finally decided to seek the expertise of several different contributors, who ended up to be more than 20, so as to provide for an adequate portrayal of the subject matter. Although I have been working on different legal topics within robotics in the past years, examining the normative challenges of such fields as the laws of war, contracts, privacy, and tortuous liability, is it wise that I now present my own book on the laws of robots? How could a single author deal with such different magnitudes of complexity, as robotics technology and the law?

There are three reasons why I believe the task is possible. First, a relatively strong consensus on how legal systems should govern the design, production and use of robots, through a complex network of concepts, such as agency, accountability, liability, burdens of proofs, responsibility, clauses of immunity, or unjust damages, still exists. In addition, jurists often claim that robotics neither creates nor modifies concepts, principles, and basic rules of the legal field, in accordance with the traditional outlook on law and robotics that may be coined here as the no new issues-thesis. In light of this popular viewpoint, one of the primary aims of this book is to test the conventional approach to the field, introducing a complex set of concepts, principles, and ways of legal reasoning pertaining to the laws of robots, in connection with Herbert H. Hart's distinction between plain and hard legal cases. As to the former set of legal issues, scholars deal with a complex web of concepts and notions in legal reasoning that yet leave no doubts as to how to apply norms and rules to a certain state of affairs, *e.g.*, cases of responsibility for the robotic behaviour pursuant to the liability model in accomplice cases of criminal law. As to the hard cases of the law, the disagreement among lawyers may regard the meaning of the terms framing the legal question, the ways such terms are related to each other in legal reasoning, or the role of the principles that are at stake in the case. Paradoxically, the fact that a strong consensus still exists in the field of the laws of robots becomes clearer when the behaviour of robots falls within the loopholes of the system, provoking a new generation of hard cases, or necessitating the intervention of lawmakers at both national and international levels. As a result, this book does not intend to offer an all-embracing depiction of today's state of the legal art and, indeed, some relevant fields such as administrative law, or crucial issues, such as data protection, are set aside. Rather, this book focuses on three legal fields, namely criminal law, contracts, and torts, so as to ascertain whether certain robotic applications, such as autonomous lethal weapons or certain types of robo-traders, truly challenge basic pillars of today's legal systems.

Second, by strictly dwelling on the legal side of robotics, instead of the physical, biological, logical, or engineering laws of the discipline, this book aims to prevent some recurring stalemates on definitional issues. Remarkably, scholars still discuss whether the behaviour of robots should properly be considered as "autonomous" and, moreover, what a robot ultimately is, namely a reprogrammable machine operating in a semi- or fully autonomous way, according to the UN World 2005 Robotics Report or, rather, a machine that can make appropriate decisions by perceiving something complex, as advocates of the "sense-think-act" paradigm propose. Such different approaches reverberate on further definitional issues as, for example, the distinction between robots and other artificial agents on the internet. Hence, in order to tackle the complexity of the field, the approach of this book is typically legal, that is pragmatic. What is at stake does not only concern the engineering meaning of such notions, as the autonomy and self-knowledge of robots, in accordance with the ways in which these machines may either interact out there with humans, and other robots, through their on-board computers, or function as robot.txt files on the web, or somewhere in between online and offline words. Rather, these notions and differences are instrumental in order to understand how these machines can affect current legal systems, much as they do, *pace* the no new issues-thesis, with crimes of negligence intertwined with matters of causation, or new kinds of responsibility for the behaviour of others in the tort law field. On this basis, the aim is to determine whether one right answer is legally at hand, whether legal systems are open to alternative solutions, or political decisions need to be taken. A typical illustration is given by the distinction between autonomous and semi-autonomous weapons in the field of military robotics, and today's debate on whether lethal force should ever be permitted to be fully automated.

Third, I concede that the time in which the intricacies of robotics technology and its impact on legal systems used to fall within the reach of a single scholar is close to an end. To date, jurists have mostly tackled the novelty of the cases induced by robotics technology with the traditional tools of hermeneutics, that is, through an extensive interpretation of the texts, through the use of analogy, the principles of the system, and so forth. In criminal law, for instance, the traditional legal viewpoint conceives robots either as dangerous animals or their use as an ultra-hazardous activity, so that strict liability rules apply to all the circumstances. In the field of contracts, rights and obligations established by artificial agents are generally interpreted through the traditional legal viewpoint of the robots-as-tools approach, so that strict liability rules govern the behaviour of these machines, binding those humans on whose behalf they act, regardless of whether such conduct was planned or envisaged. In tort law, strict liability rules in the field of robotics are most of the time understood by analogy with a party's responsibility for the behaviour of animals, children, or even employees. Yet, the more robotics advances and becomes more sophisticated, the more likely it is that such machines will need a legal regime of their own. Among the solutions proposed in this book, contemplate new forms of accountability for the behaviour of robots in the field of contracts, which mean that, under certain circumstances, only robots would be held liable for damages caused by them. Likewise, consider new forms of responsibility for the behaviour of others, *e.g.*, robots in the field of torts, so that clauses of negligence-based responsibility could replace some of today's strict liability rules in cases where third parties are the least-cost avoider of the risk. At the end of the day, the aim of this work is not only to pinpoint those principles, norms, and concepts of today's legal systems which are under stress: the purpose is also to take sides before the hard cases of the law as induced by a novel generation of robotic applications. All in all, I think that some types of robots should not be considered as simple tools of human interaction but, rather, proper agents in the legal field.

However, the more robots require a legislation of their own, the more a new team of experts in robotic crimes, pacts and contracts, administrative procedures, copyright and privacy issues, laws of war, torts, and so on, will supersede the efforts of the single scholar. The process of specialization that has occurred within such fields as IT law, or legal informatics, throughout the 2000s, will likely resurface in the field of legal robotics in a few years. Retrospectively, this work is placed at a turning point of the contemporary legal systems, that is, so to speak, between a "not yet" and an "any longer." Not yet, because a number of challenges brought on by robotic technology and its manifold applications are still open to alternative solutions in the legal domain; any longer, because traditional legal outlooks increasingly fall short in coping with the novelty of such challenges. Let us grasp why we are facing such an in-between state of art in the laws of robots, throughout the chapters of this volume.

Torino, Italy Ugo Pagallo

Acknowledgments

This book is the final step of a 4-year project (2009–2013). The preliminary stages were the papers and articles that I have been discussing and publishing over the past years. First, thanks are given to the referees and editors of the journals and books where I published my previous robotic works, a detailed list of which is given below in the references. In particular, let me thank Mariarosaria Taddeo, who edited the special issue of *Knowledge, Technology & Policy* (2010: 23) on "Trust in Technology," in which I presented "Robotrust and Legal Responsibility" (Pagallo 2010a); Terry Bynum and Simon Rogerson, the founders and souls of the Ethicomp meetings, where I delivered "The Human Master with a Modern Slave?" (Pagallo 2010b), and "The Adventures of Picciotto Roboto" (Pagallo 2011a); Greg Michaelson and Ruth Aylett, editors of the special issue of *AI & Society* (2011: 26(4)) on the "Social Impact of AI: Killer Robots or Friendly Fridges," with my "Killers, Fridges, and Slaves" (Pagallo 2011b); John Sullins, who edited the special issue of *Philosophy & Technology* (2011: 24(3)) on "Open Questions in Roboethics," with my "Robots of Just War" (Pagallo 2011c); Herman Tavani, editor with Dieter Arnold of the special issue of *Information* (2011: 2(2)) on "Trust and Privacy in Our Networked World," where "Designing Data Protection Safeguards Ethically" was published (Pagallo 2011d); Brendan Gogarty, who invited me to deliver an expert commentary for the special edition of the *Journal of Law, Information and Science* (2011) on "Laws Unmanned," that is, my paper on "Guns, Ships, and Chauffeurs" (Pagallo 2011e); and, last but not least, Mireille Hildebrandt, who edited with Jeanne Gaakeer the Springer volume on "Human Law and Computer Law," with my essay on "What Robots Want" (Pagallo 2013).

All this previous work represents the starting blocks of this volume, together with both the papers for the AICOL series, coedited with Gianmaria

Ajani, Pompeu Casanovas, Monica Palmirani, and Giovanni Sartor (Pagallo 2010c, 2012a); and the entry "Robotica" for the UTET volume on legal informatics, coedited with Massimo Durante (Pagallo 2012b). During the Fall semester of 2011, spent at the University of Uppsala, a first draft of this book was completed thanks to the formal revision and substantial remarks of Patricia Mindus and Laura Carlson. The manuscript was revised a second time during the Spring semester of 2012, spent at my own university in Turin, where I delivered my course on legal informatics and robotics. A number of colleagues and friends should be thanked for their support, much as the students of my course for their questions and theoretical curiosity. Together with Gianmaria Ajani and Massimo Durante, let me especially thank Raffaele Caterina and Michele Graziadei. By the end of April 2012, I then followed the advice of Greg Chaitin: after pinning the book down as slowly as possible, I let it rest for a while. Three months later, in August 2012, a third revision was undertaken and the preface completed in Cupertino, CA. During the wonderful weeks spent in my favourite villa, I enjoyed the further insights of an eminent expert in machine learning and AI, namely my sister Giulia, and of a distinguished mathematician, my brother-in-law Victor Pereyra.

Diachronically, the book was also improved or, at least, some of its limits and vagueness superseded thanks to many conversations with Luciano Floridi, with whom I had the honour to be member of the group of experts on "the onlife initiative," set up by the European Commission as part of the Digital Futures project in 2012. The final revision of the book was completed in January 2013, paying attention to the remarks of the reviewers, much as the suggestions of further colleagues and friends. Among them, let me mention Chuck Abernathy from Georgetown University for his common law wisdom. As to the practical side of this volume, thanks are given to the editors of the Springer series on "Law, Governance and Technology," that is Pompeu Casanovas and Giovanni Sartor, together with Neil Olivier, Senior Publishing Editor of Springer, and his assistant Diana Nijenhuijzen. From the initial project of the book in August 2011 to the green light of the Springer team by the end of winter 2012–2013, all of them helped me to make that August' 11 project real.

Despite this manner of production and the number of inputs by reviewers, colleagues, and friends, I am conscious that the book may still have ambiguities, imprecisions, or simply mistakes. This possibility reminds me of the introductory scene of Ĉapek's *Rossum's Universal Robots*, where Domin, the General Manager of R.U.R., explains to Helena that they were building robots by the thousands, able to speak, write and do arithmetic, free from errors and with a formidable memory. This original idea has significantly fed popular beliefs ever since, down to the point that a pop song recently reminds us that "I am not a robot." Although one of the most critical issues of robotics concerns the degree of their error-proneness, let alone whether these machines could have some types of emotions, like falling in love, this naïf version of the field functions as a proemial warning. The continuous process of reviewing and the suggestions of colleagues and friends helped me to improve the previous versions of this book and, yet, some imperfections may still remain. Updating Augustin of Hippo's proverb, to err is human, to persist is of the bad robotic designer.

Contents

1 Introduction 1

2 On Law, Philosophy and Technology 19

2.1 The Philosophy of Law and Robots 21

2.1.1 The Law in Literature 22

2.1.2 Sources, Concepts, and Legal Reasoning 25

2.1.3 The Levels of Abstraction 28

2.2 The Principle of Responsibility 29

2.2.1 Immunity 31

2.2.2 Strict Liability 33

2.2.3 Personal Fault 34

2.2.4 Responsibility for a Robot 35

2.3 Agency and Accountability of Artificial Agents 37

2.3.1 A Moral Threshold 38

2.3.2 Agents Before the Law 40

2.4 Who Pays? 43

3 Crimes 45

3.1 Sci-Fi Scenarios 49

3.2 The States of Mind and Criminal Acts 52

3.3 Robots and Just Wars 55

3.3.1 What Robots Might Change 57

3.3.2 Just Causes of War 58

3.3.3 Conditions of Just Wars 60

3.3.4 Proportionality 62

3.4 The Phenomenology of *Picciotto Roboto* 65

3.4.1 Picciotto by Design 66

3.4.2 Crimes of Intent 69

3.4.3 Crimes of Negligence 71

3.5 A Failure of Causation? 73

4 Contracts ... 79
4.1 Pacts, Clauses and Risk ... 83
4.2 The Artificial Doctor ... 88
4.2.1 Parties, Counterparties and Third Parties ... 89
4.2.2 Producers, Users and Patients ... 91
4.3 Robo-Traders ... 95
4.3.1 Artificial Greediness ... 96
4.3.2 The Robot and the Principal ... 97
4.3.3 A New Agent in Town ... 101
4.4 Modern Robots, Ancient Slaves ... 102
4.4.1 The Digital Peculium ... 103
4.5 The UV Revolution ... 106
4.5.1 AI Chauffeurs and Intelligent Car Sharing ... 108
4.5.2 Unjust Damages ... 111

5 Torts ... 115
5.1 Bad Intentions ... 119
5.2 Children, Pets and Negligence ... 121
5.2.1 American Parents ... 124
5.2.2 Italian Parents ... 126
5.3 AI Employees and Strict Liability Rules ... 130
5.3.1 The Digital Peculium Revisited ... 132
5.4 Burdens of Proof ... 135
5.4.1 The Precautionary Principle ... 138
5.4.2 Robotic Openness ... 143

6 Law as Meta-technology ... 147
6.1 Robots as Legal Persons ... 152
6.1.1 The Front of Robotic Liberation ... 155
6.1.2 The Pragmatic Stance ... 163
6.2 Robots as Strict Agents ... 166
6.3 Sources of Good and Evil ... 170
6.4 Levels of Complexity ... 174
6.4.1 Technologies of Social Control ... 177
6.4.2 The Political Requirement ... 179

Conclusions ... 183

References ... 193

List of Figures

Fig. 1.1 The magnitudes of complexity of robotics technology 7
Fig. 1.2 A philosophy of law for lawyers and a work in positive law for philosophers 10
Fig. 1.3 Three legal fields for responsible robots 14

Fig. 2.1 Levels of abstraction 22
Fig. 2.2 A first model for the philosophy of law and robots 25
Fig. 2.3 A second model for the philosophy of law and robots 27
Fig. 2.4 A new interface for the philosophy of law and robots 28
Fig. 2.5 Three conditions of responsibility for the construction and use of robots 31
Fig. 2.6 From responsibility to legal agency and return 41

Fig. 3.1 The Phenomenology of Picciotto Roboto, step 1 67
Fig. 3.2 Phenomenology of Picciotto Roboto, step 2 70
Fig. 3.3 Phenomenology of Picciotto Roboto, step 3 72

Fig. 4.1 Contractual obligations and robotics complexity 81

Fig. 5.1 A common law approach to negligence in the law of Torts 125
Fig. 5.2 A civil law approach to the law of Torts 127
Fig. 5.3 Strict liability for robots in the law of Torts 130
Fig. 5.4 Reversing the burden of proof with the precautionary principle 139

Fig. 6.1 Law and the challenges of technology 149
Fig. 6.2 Levels of legal complexity in the governance of robotics 175
Fig. 6.3 Four robotic challenges to law as meta-technology 178

Fig. A.1 Three roads to design 184
Fig. A.2 A teleological approach to design 186

List of Tables

Table 1.1 The behaviour of robots and nine ideal-typical conditions of legal responsibility 13

Table 4.1 What the approach to robots-as-tools lacks 100

Table 6.1 Robots' behaviour and the "Factual Limits" of legal science 165

Table 6.2 A threshold of robots' responsibility in the civil law field 169

Table 6.3 The challenges of today's laws of robots as a source of damage 173

Chapter 1
Introduction

HELENA:	*You mean you make them start to work as soon as they're made?*
DOMIN:	*Sorry. It's more like working in the way a new piece of furniture works…*
HELENA:	*How do you mean?*
DOMIN:	*Much the same as going to school for a person. They learn to speak, write, and do arithmetic. They have a phenomenal memory. If one reads them a twenty-volume encyclopaedia, they could repeat it back to you word for word, but they never think up anything original. They'd make fine university professors.*

Karel Ĉapek, *Rossum's Universal Robots,*
Introductory Scene

Abstract The aim of this book is to introduce laypersons to the complex set of principles, concepts, and ways of legal reasoning that govern the design, construction, supply and use of robotics technology today. In light of the classical distinction between legal plain and legal hard cases, attention is drawn to the cases where the disagreement among lawyers regards either the meaning of the terms framing the legal question, or the ways such terms are related to each other in legal reasoning, or the role of the principles that are at stake in the case. Paradoxically, the fact that a strong consensus still exists in the field of the laws of robots becomes clearer when the behaviour of robots falls within the loopholes of the system, provoking a new generation of hard cases.

U. Pagallo, *The Laws of Robots: Crimes, Contracts, and Torts*, Law, Governance and Technology Series 10, DOI 10.1007/978-94-007-6564-1_1,

The two different magnitudes of complexity as explored in this book, robotics technology and the law, challenge not only each other, but also today's society. Following the term Isaac Asimov coined in his 1942 novel, *Runaround*, "robotics" is the field dealing with the design and construction of a quantity of machines as varied as network centric-applications, adaptive robot servants, robot soldiers, unmanned ground and underwater vehicles, robot toys and even robot nannies. Robotics today is one of the most exciting fields of scientific research and technology, spanning several disciplines, such as artificial intelligence ("AI") and computer science, cybernetics, physics and mathematics, electronics and mechanics, neuroscience, biology and the humanities. Despite the multiplicity of robotic applications, some argue that we are dealing with machines built basically upon the mainstream "sense-think-act" paradigm of AI research (Bekey 2005). Sebastian Thrun, director of the AI Laboratory at Stanford, California, similarly reckons that robots are machines with the ability to "perceive something complex and make appropriate decisions" (in Singer 2009: 77). Others stress that robots are those machines able to learn and adapt to changes in environments. The UN World 2005 Robotics Report proposes a general definition of robot as a reprogrammable machine operating in a semi- or fully autonomous way, so as to perform manufacturing operations (*e.g.*, industrial robots), or provide "services useful to the well-being of humans" (*e.g.*, service robots).

These definitions do not dispel all doubts. References to the autonomy or intelligence of robots often are a source of misunderstanding. Consider the UK Ministry of Defence's Joint Doctrine Note on "unmanned aircraft systems" dated 30 March 2011. The notion of autonomy there is connected to a system "capable of understanding higher level intent and direction." Moreover, according to the Note, "estimates of when artificial intelligence will be achieved (as opposed to complex and clever automated systems) vary, but the consensus seems to lie between more than 5 years and less than 15 years, with some outliers far later than this." Opponents find this statement "ludicrous": in *Automating Warfare* (2011), Noel Sharkey affirms that, apart from the metaphorical use of the words, robots are not going to be "capable of understanding higher level intent," nor will they think like human beings in the foreseeable future. Likewise, Kenneth Himma argues in *Artificial Agency* (2007) that robots and other artificial agents ("AAs") do not meet the necessary and sufficient conditions required for properly claiming they engage in autonomous behaviour, as AAs lack the requisites of consciousness, free will and intent.

Sci-Fi scenarios aside, certain types of robots are already challenging tenets of social interaction, basic rules among nations, and even cornerstones of the law. "Even if they have the intelligence of a refrigerator"

(Floridi 2007), robots can improve the set of instructions through which their inner states change, and transform such properties without external stimuli: therefore, they can deal successfully with their tasks by exerting control over their own actions without any direct intervention by humans. As the 2007 EURON Roboethics Roadmap states, "in a few years we are going to cohabit with robots endowed with self-knowledge and autonomy – in the engineering meaning of these words" (Veruggio 2006). This specific autonomy of the robot, taking decisions of its own, seems particularly critical in such fields as military robotic technology: the United States military forces fund more than one-half of the American research and development ("R&D") in AI today. Consequently, looking at certain military robotic applications is instructive in shedding further light on the notion of robots that can rule (*nomos*) over themselves (*auto*) and, thus, are autonomous in a general sense.

For example, in the field of unmanned aerial vehicles ("UAVs"), a distinction should be drawn between "autonomous" and "semi-autonomous" machines. Some drones, such as the US Air Force's RQ-1 and MQ-1 Predators, have to be considered semi-autonomous. Others are fully "independent of real time UAV-pilot control input," according to the UK Defence Standards definition of autonomous flight. Think of the Global Hawk and the US Navy's anti-ship missile defence system, the Phalanx CIWS, operating completely alone. Some 40 countries currently are developing even more sophisticated forms of autonomous lethal weapons and other types of robot soldiers, a development summed up by scholars as "killer robots" (Sparrow 2007; Krishnan 2009), "robotic lethal behaviour" (Arkin 2007), or "autonomous military robotics" (Lin et al. 2008). Although these machines are not conscious of themselves and do not enjoy any "higher level intent and direction," they can act and decide beyond the direct control of humans. Norbert Wiener justly warned about the "autonomy of robots" in *The Human Use of Human Beings* (1950): the use of robots in battle might lower the requirements of declaring or entering into war, invoke a disproportionate use of force, violate the principle of discrimination and immunity, and might even provoke accidental wars. By considering the impact of today's robot soldiers on traditional categories of *ius ad bellum* (*i.e.*, when and how resort to war can be justified) and *ius in bello* (*i.e.*, what can justly be done in war), it can be remarked that the menace of robotic behaviour is as old as the very idea of "robot."

The word "robot" was used for the first time in Karel Ĉapek's 1920 play, *Rossum's Universal Robots*. The plot revolves around a factory producing artificial persons, "robots," whose rebellion ultimately leads to the extinction of the human race. In the second act, individuals at the headquarters of

R.U.R., the world manufacturer of thousands of robots located on a remote island, wonder why these machines are revolting against humanity. Dr. Gall, Head of the Physiology and Research Department at R.U.R., reckons that the "crucial mistake" they made was to turn some of these machines into "robot soldiers."

> This is just the same old evil as Europe has always committed. They just couldn't leave their damned politics alone and so they taught the robots to go to war, they took the robots and turned them into soldiers and that was a crime against humanity (Ĉapek 1920, Act 2).

Reality, at times, outpaces fantasy: since 2005, combat air patrols by US drones have increased by 1,200 % and, under President Barack Obama, the frequency of such strikes in Pakistan has risen tenfold "from one every 40 days during George Bush's presidency to one every four" (*The Economist*, 8 October 2011, p. 32). Significantly, Christof Heyns, Special Rapporteur on extrajudicial executions, urged in his 2010 Report to the UN General Assembly that Secretary-General Ban Ki-moon convene a group of experts in order to address "the fundamental question of whether lethal force should ever be permitted to be fully automated."

Robotics behaviour has appeared to be a source of risk and potential threat in other realms as well: the financial troubles in late 2008 may have been facilitated by the use of "robo-traders" such as AI brokers, electronic agents and smart digital interfaces. Since the early 2000s, experiments with Zero Intelligent ("ZI") agents, developed by the University of Pennsylvania and Lehman Brothers, have shown troubling similarities to the greediness of human speculators. In *Rights of Non Humans?* (2007), Günther Teubner sums up these concerns, claiming that robotics technology and other smart artificial agents raise problems of alienation and reification in social life that already troubled Karl Marx (*Entfremdung*) and Martin Heidegger (*Verdinglichung*). The overall idea is that autonomous AAs "create aggressive new action centres as basic productive institutions" so that we should bring the "economic, social and technical transactions run by electronic agents... back under human control" (Teubner 2007: 21).

Admittedly, the use of robo-traders in financial markets and autonomous lethal weapons on battlefields is alarming. However, let us avoid sweeping generalizations. Rather than machines that necessarily "alienate" (Marx) or "reify" (Heidegger) human life, we should pay attention to the number of robotics applications that, according to the UN World 2005 Report, provide "services useful to the well-being of humans." To start with, think of intelligent vehicles driving themselves on highways, a popular subject of Sci-Fi movies such as Michael Keaton's Batmobile in *Batman* (1989), or, for that matter, the smarter AI cars in *Demolition Man* (1992), *Timecop* (1993), *Minority Report* (2002) and *I, Robot* (2004). Over the past decade, research (*e.g.,* Stanford University and Carnegie Mellon), business (General Motors and Volkswagen), and both (Google), have made this dream come true. To cut to the chase, the Nevada Governor in June 2011 signed a bill into law that for the first time ever authorizes the use of driverless cars on public roads. Of course, this is not to say that today's AI chauffeurs are as sophisticated as the Sci-Fi cars in Hollywood movies. Moreover, the Nevada Assembly (36–6) and Senate (20–1) acknowledged that "regulations authorizing the operation of autonomous vehicles on highways within the State of Nevada" may take a long time. Still, *pace* Teubner, robotic automation might not be a bad thing, once we recall that the autonomy of human drivers causes around 1.3 million accidents and 41,000 deaths on EU roads every year.

Likewise, contemplate certain useful applications in the industrial and service sectors. For example, a new generation of unmanned water-surface and underwater vehicles for remote exploration began in the 1990s to undertake emergency and hazard management work, by preventing damage, alerting controllers, fixing oil leaks, and so forth. Some of these underwater robots became popular in 2010, when they were employed for stopping the BP oil spill in the Caribbean Sea. In addition, a number of artificial companions and helpers at home, such as robot toys and robot nannies, are programmed in the field of service robots for domestic or personal use, to provide love and take care of children and the elderly. In the show business and music industry, consider the success story of the Japanese pop star robot singer HRP-4C. Developed by the Institute of Advanced Industrial Science and Technology's media interaction group, this amazing "divabot" is capable of singing, dancing, "breathing," and even performing her (!) shows. While HRP-4C uses the Vocaloid software developed by Yamaha, as well as a VocaListener to synthetize the notes of the songs, a VocaWatcher program allows HRP-4C to analyse individuals' facial tics as this divabot moves her hips and belts out a tune. Although it may be conceded that a robotic Maria Callas would be more stimulating than the current robotic Lady Gaga, it is difficult to see why this machine should *a priori* be likened to her more troubling cousins, robo-traders and robot soldiers. Some of these AI nannies and show biz pop girls raise a number of psychological issues concerning feelings of subordination, attachment, trustworthiness, etc. Yet, going back to some current picture of robotics, *e.g.*, Teubner's *Rights of Non Humans?*, it is problematic to dismiss such robots as an expression of "aggressive new action centres."

Robotic applications bring about a new set of constraints and opportunities that transform, reshape and even enrich individual and social environments.

This twofold aspect of robotics, as a source of good and evil, has been stressed by a number of scholars who, interestingly, insist on the impact of both military robotics technology and the service robots "useful to the well-being of humans," illustrated by the UN 2005 Report. Introducing a special issue of *AI & Society* on "the social impact of AI" (2011), Greg Michaelson and Ruth Aylett emphasize that "the recent advancements in the now mature discipline of Artificial Intelligence… have rekindled problematic social and ethical questions about our relationships with machines," adding to the tension between "killer robots" and "friendly fridges." Similarly, in the introduction to the special issue of *Philosophy & Technology* on "robotics: war and peace" (2011), John Sullins reckons that the ethical questions about our relationships to robots can be fruitfully addressed in connection with the following spectrum: at one end, "robots of war" such as MQ-9 Reapers or C-3PO Terminators may be presented as emblems of the "aggressive new action centres" of Teubner's version of robotics; at the other end of the spectrum are "robots of peace," such as the Japanese pop singer HRP-4C or, say, the da Vinci surgery system in the medical sector. What is common to robotics, from this point of view, ultimately revolves around the normative challenges of this technology, that is, "why we should, or should not, deploy these systems in our homes and battlefields" (Sullins 2011).

The kinds of robotic applications we are willing to implement is a crucial question today for ethics, economics, philosophy of technology, psychology and other fields. Here, the focus is not on how the manifold applications of robotics technology obey the "laws" of disciplines such as mathematics, physics, neuroscience, biology, and so forth. Rather, attention is drawn to the reasons why such machines should, or should not, be deployed in accordance with the aim of the moral, political and economic fields, in governing the process of technological innovation. Figure 1.1 below shows how the different magnitudes of complexity concerning the "laws of robots" can be illustrated:

Let us now augment the intricacy of this model by focusing on the second magnitude of complexity as explored in this book. In addition to multiple robotic applications and the laws of such disciplines as AI and computer science, cybernetics, and so on, that which is under scrutiny concerns the legal challenges facing this field: "the laws of robots." The first problem is identifying what is common to robotics through the lens of the "laws of the law."

Traditionally, when determining "what the law is," scholars distinguish the law from other academic fields, such as politics, ethics or economics. However, certain scholars affirm that the law ultimately depends on such

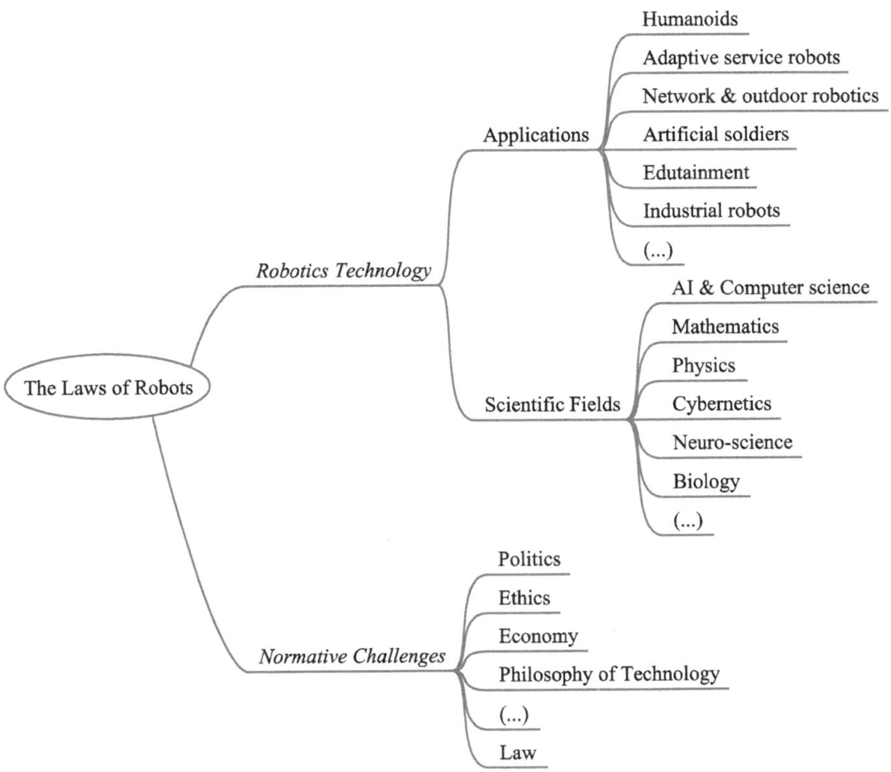

Fig. 1.1 The magnitudes of complexity of robotics technology

fields: a realist would trace the law back to politics, an advocate of the natural law tradition to ethics, an expert of the economic analysis of law (as well as an orthodox Marxist) to economics, a techno-determinist scholar to technology, and so forth. It suffices to mention the thesis of a "reductionist," such as the Italian philosopher Benedetto Croce. In *Riduzione della filosofia del diritto alla filosofia dell'economia* (1907), Croce sums up the efforts of legal philosophers to distinguish their own field of law from morals, by the image of the "Cape Horn" of legal science. The overall idea is that lawyers, trying to circumnavigate this issue, end up in a "conceptual storm" and "wreckage." In light of today's debate in legal theory, and how the variations of positivism (both inclusive and exclusive), realism, institutionalism, and different traditions of natural law, perceive the connection between law and morals, some words on the normative fabric of the legal phenomenon seem necessary, in order to clarify the legal approach of this book to the laws of robots. The nature of law and its connection with the moral sphere can be properly understood by examining circumstances under which individuals (and robots) are confronted with responsibility.[1]

Reflect on cases where responsibility is imposed on individuals for harm resulting from their own fault. This is typical when an individual voluntarily performs a wrong prohibited by law, *e.g.*, tiny robotic helicopters employed in a jewellery heist. In criminal law, the legal accountability for this kind of behaviour is entwined with the notion of the moral responsibility of the individual and the idea of blameworthiness. Criminal defendants ought to be subject to the ordinary process of moral assessment in order to determine whether they are guilty under the law. In civil (as opposed to criminal) law, the general idea is similar, in that individuals are held liable for unlawful or accidental damages caused to others due to personal fault. This idea is traditionally summed up by the Roman maxim, *alterum non laedere*, that is, "do not injure others." Although further examples can be given, it should be clear that legal and moral reasons can overlap. We return to this below.

However, there are other circumstances in which individuals find themselves confronted with legal responsibility and yet, the actor's moral responsibility is not at stake. The first case of legal (as opposed to moral) responsibility refers to the idea that "everything which is not prohibited is allowed." In criminal law, this principle is connected to the clause of immunity summed up, in continental Europe, with the formula of the principle of legality, *i.e.*, "no crime, nor punishment without a criminal law" (*nullum crimen nulla poena sine lege*). Even though certain behaviours might be deemed as morally wrong, *e.g.*, spying on individuals through domestic robots, individuals can be held criminally liable for that behaviour only on the basis of an explicit criminal norm. In the wording of Article 7 of the 1950 European Convention on Human Rights, "[n]o one shall be held guilty of any criminal offence on account of any act or omission which did not constitute a criminal offence under national or international law at the time when it was committed."[2] *Vice versa*, there are cases where the law establishes no-fault liability, that is, regardless of the person's intent or ordinary care. Although a conduct may be deemed morally sound, a statute or a specific norm can establish liability for that behaviour. An example of this can

---

[1]The connection between the law and such fields as politics, economy, and technology, is further examined in Chap. 5.

[2]As lawyers know, there is a savings provision pursuant to art. 7(2) of the Convention, which states: "This article shall not prejudice the trial and punishment of any person for any act or omission which, at the time when it was committed, was criminal according the general principles of law recognized by civilized nations." The aim of this provision is to cover such exceptional cases as the Nuremberg trial against the Nazis.

be seen with editors, publishers and media owners (newspapers, TV channels, radio, etc.), parties who are liable for damages caused by their employees, notwithstanding their eventual illicit or culpable behaviour. This mechanism is invoked in many other types of cases where the law imposes liability regardless of the person's intention. Besides individuals' responsibility for the behaviour of their pets and, in most legal systems, their children, this type of strict liability applies to most producers and users of robots.

Going back to Croce's Cape Horn in legal theory, we can shed further light on the normative efforts of the law from a broader perspective, that is, by distinguishing plain from hard cases (*e.g.*, Hart 1961; and Dworkin 1986). A way to circumnavigate Croce's problem exists: we can avoid storms and conceptual wreckages by drawing attention to all the cases where a complex set of concepts and notions in legal reasoning are at work and, still, leave no doubts as to how to apply the clauses and conditions of responsibility/liability in the legal field. According to Herbert Hart, these are the cases where the legal issues are pretty plain, that is, "where the general terms seem to need no interpretation and where the recognition of instances seems unproblematic or 'automatic'… where there is general agreement in judgements as to the applicability of the classifying terms" (Hart 1994: 123). Clauses of immunity in criminal law and cases of no-fault liability in tort law may thus represent a class of such plain cases, in that, here, the distinction between an individual's moral and legal responsibility is not an issue at all. Throughout this book, we are going to see further examples of this general agreement on how the principles, norms, and rules of the legal system work: namely, cases of responsibility pursuant to the liability model in accomplice cases of criminal law (Chap. 3), cases of responsibility that depend on the voluntary agreement between private persons in the civil law field (Chap. 4), down to the strict liability hinging on the idea of dangerous activities in tort law (Chap. 5). This network of concepts in legal reasoning allows scholars to examine matters of unpredictability and risk as provoked by robots, as was the case with previous technological innovations.

Still, there are cases where scholars (and parties to a lawsuit) may disagree. Here, the storms and conceptual wreckages of Croce's Cape Horn represent a class of legal issues that scholars dub as hard cases, for instance, where the disagreement may regard the meaning of the terms that frame the legal question, or the ways such terms are related to each other in legal reasoning, or the role of the principles that are at stake in the case. However, which principles, which concepts, and which ways of legal reasoning, at times end up in a sort of legal stalemate has to be determined in connection with the norms and provisions established by statutes, international agreements, or the case law of the common (as opposed to the civil) law tradition. Work on the logic and nature of the law, such as Croce's own research in legal philosophy, in other words is

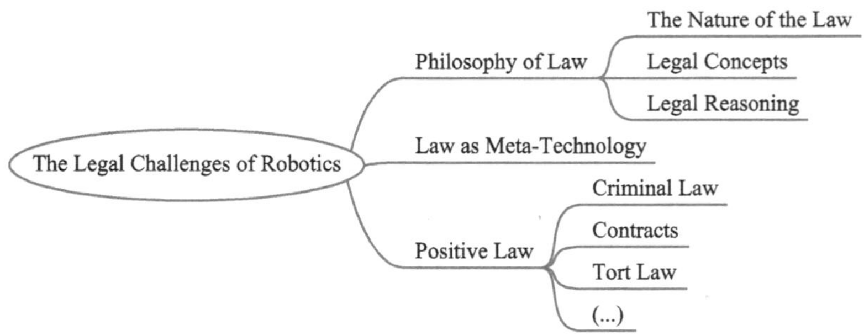

Fig. 1.2 A philosophy of law for lawyers and a work in positive law for philosophers

a necessary, but insufficient ingredient of the analysis: in order to determine whether a legal issue appears hard, or plain, we need the knowledge of experts in positive law as much as the efforts of legal philosophers. For example, regarding the military employment of robotic applications, focus should be on the 1907 Hague Convention, the four Geneva Conventions from 1949, and the two 1977 additional Protocols, which define the current laws of war and the international framework of humanitarian law. In the case of, say, the civilian use of unmanned aerial vehicles, attention should be drawn to the 1948 Chicago Convention on International Civil Aviation and, in Europe, the EU Regulation 216/2008. In the case of the civilian use of unmanned water-surface and underwater vehicles, the legal point of reference is the 1972 IMO COLREGs Convention on maritime law.

This twofold approach to the laws of robots, that is, both the perspective of legal philosophers and the knowledge of experts in positive law, can be summed up with a sort of interface, or level of abstraction,[3] through which this book aims to describe, examine, and argue about the laws of robots. What I propose here is to approach the laws of the law establishing the conditions of legitimacy for the design, production, and use of robots, conceiving the law as meta-technology, *i.e.*, as a means to govern other technological means. This perspective sheds further light on topics of legal philosophy (*e.g.*, the nature of the law, concepts, legal reasoning), as well as provisions of positive law. Figure 1.2 sums up this level of abstraction:

As seen from Book IV of Plato's *The Republic*, this idea is not new: "The regulations which we are prescribing, my good Adeimantus, are not, as might

[3] On the methodology of the "level of abstraction," this author draws on Luciano Floridi's work. See *The Method of Levels of Abstraction* (2008) and, more recently, the second volume of Floridi's *Principia Philosophiae Informationis*, namely Information Ethics (2013). By varying the "interface," the "set of observables" changes accordingly: more details on this method in Sect. 2.1.3.

be supposed, a number of great principles, but trifles (Plato 2006)." In this context, the regulative efforts of the law can be illustrated with the thesis of the *Pure Theory of Law* (1934/2002) and *General Theory of the Law and the State* (1945/1949). Here, Hans Kelsen provides a classical account of the law as "a specific social technique of a coercive order" enforced through the menace of physical sanctions: "if A, then B." The legal formula shows "what should be" (*Sollen*, ought to), rather than "what is" (*Sein*, is), namely, punitive sanctions (B) that should follow terms and conditions of legal accountability (A), rather than effects (B) that follow natural causes (A). The distinction between normativity and natural causality means that the aim of the law, to govern the conditions of legitimacy for technological innovation (A), hinges on what should happen in terms of legal responsibility (B). In the phrasing of the *General Theory of the Law and the State* (1949: 26): "What distinguishes the legal order from all other social orders is the fact that it regulates human behaviour by means of a specific technique." Once such technique regulates other techniques and, moreover, the process of technological innovation, we may accordingly conceive the law as a meta-technology.

To be sure, law can be considered as a form of meta-technology without buying Kelsen's ontological commitment. The stance this book adopts does not imply either that the law is merely a means of social control, or that there are no other meta-technological mechanisms. Rather, the level of abstraction defined by law as meta-technology aims, first, to describe how legal systems deal with the process of technological innovation, through such a complex network of concepts, as agency, accountability, liability, burdens of proofs, clauses of immunity, or unjust damages. The analysis dwells on the conditions of legitimacy for the design, construction, and use of robots, as scholars have done since they started examining the impact of automation on the law in the late nineteenth century. Think of Günther's *Das Automatenrecht* (1892), Schels' *Der strafrechtliche Schutz des Automen* (1897), Schiller's *Rechtsverhältinesse des Automen* and Ertel's *Der Automatenmissbrauch und seine Charakterisierung als Delikt*, both from 1898, to Neumond's *Der Automat* in 1899. More than a century later, there is still a relatively strong consensus: in a great number of cases, the rules that govern the design, production and use of such machines (Kelsen's A) are unchallenged, as well as the consequences in terms of legal responsibility (B).

Then, *pace* Kelsen, we should pay attention to the impact of robotics technology on the formalisms of the law, and how we grasp the meaning of certain key terms concerning the aim of the law to govern the process of technological innovation. This impact brings us back to the hard cases of the law, and how we should address them. Some affirm "there is no possibility of treating the question raised by the various cases as if there were one uniquely correct answer to be found, as distinct from an answer which is a reasonable compromise between many conflicting interests" (Hart 1961: 128). Others, as Ronald Dworkin and followers of the "right answer" thesis, on the contrary interpret the law in a morally coherent way, so that, given the nature of the legal question and the history and background of the issue, *e.g.*, whether to ban robot soldiers through a UN sponsored agreement, lawyers could obtain the solution that best justifies or fits the integrity of the law.

That suggested here is restricting the focus of the analysis and summarizing the complex set of principles, norms and rules establishing the conditions of legitimacy for the design, production and use of robots, through the concepts of legal responsibility (Kelsen's B) and agency (*i.e.*, a key term of Kelsen's A). This stricter perspective emphasizes that which all cases concerning the laws of robots have in common, namely, the conditions whereby legal agents, both human and artificial, are confronted with responsibility. Whether a unique right answer exists (*e.g.*, Dworkin), or not (Hart), we have to preliminarily ascertain the terms through which the law frames technological research and development, so as to take sides in today's debate. Theoretically speaking, three legal notions of agenthood are at stake:

(i) Legal persons with rights (and duties) of their own;
(ii) Proper agents establishing rights and obligations in civil law;
(iii) Sources of responsibility for other agents in the system.

Likewise, the different types of cases where agents are confronted with legal responsibility should be stressed:

(i) The aforementioned clauses of immunity (*e.g.*, the principle of legality);
(ii) Conditions of strict liability (*e.g.*, no-fault responsibility of editors);
(iii) Cases of responsibility for damages that depend on fault (*e.g.*, intentional torts).

On this basis, three different levels of analysis can be distinguished:

(i) The different ways robots do act in legal systems (Kelsen's A);
(ii) The consequences following from the production and use of such machines (Kelsen's B);
(iii) The overall impact of technology on legal systems, so as to determine whether a case is plain, or hard (*e.g.*, Dworkin vs. Hart).

Table 1.1 summarizes this approach with nine possible scenarios:

The legal observables of responsibility for the behaviour of robots in light of Table 1.1 clarify the philosophical challenges of the field, *e.g.*, its hard cases, and the matters of responsibility in positive law, *e.g.*, robotic crimes.

Table 1.1 The behaviour of robots and nine ideal-typical conditions of legal responsibility

Responsible robot	*Immunity*	*Strict liability*	*Unjust damages*
As legal person	I-1	SL-1	UD-1
As proper agent	I-2	SL-2	UD-2
As source of damage	I-3	SL-3	UD-3

Let us become acquainted with such ideal-typical conditions of responsibility for the behaviours of robots:

"I-1," "SL-1," and "UD-1" have in common that robots should be considered as proper persons with rights (and duties) of their own, that is, the thesis of what I call the front of Robotic Liberation. "I-1" means that a person is protected by clauses of immunity, *e.g.*, the principle of legality. "SL-1" stands for cases of no-fault responsibility of the robot as being *sui iuris*. Finally, "UD-1" concerns protection against harm provoked by others: for example, the State, contractual counterparties, third parties in tort law.

"I-2," "SL-2," and "UD-2" share the idea that (some types of) robots can properly be conceived as strict agents in business law: for example, with negotiations and contracts. "I-2" has to do with clauses of immunity in the civil (as opposed to the criminal) law field, such as protection pursuant to safe harbour clauses. "SL-2" *vice versa* emphasizes liability of this robot agent, regardless of intentions or personal fault. Then, "UD-2" stresses that such agents should be protected against unjust damages.

Finally, "I-3," "SL-3," and "UD-3" summarize the traditional viewpoint of scholars that robots would not affect basic cornerstones of the law. As simple tools, and not agents, in the legal system, robots can only represent a source of responsibility for other agents. Therefore, "I-3" means that humans, as well as artificial persons such as corporations, evade responsibility for damage provoked by robots, *e.g.*, clauses of immunity in the laws of war. "SL-3" highlights today's strict liability policies for the design, construction and use of robots. "UD-3" concerns cases of responsibility for human negligence or intentional wrongdoing, which have to be added to the previous hypothesis of no-fault responsibility.

In light of Table 1.1, the complex network of concepts, through which the law aims to govern the process of technological innovation, results in the traditional focus on the question of "Who pays?" This question suggests three scenarios for a hard case in positive law. The disagreement can concern:

(i) The legal personhood of robots and their constitutional rights;
(ii) The legal accountability of robots in contracts and how this autonomy impacts other fields of the law;
(iii) New types of human responsibility for others' behaviour.

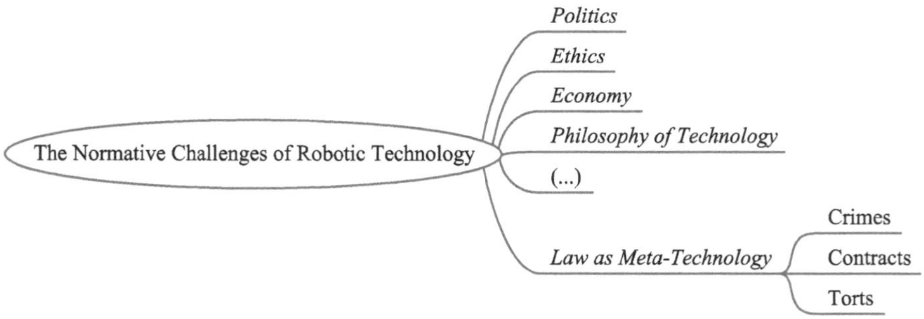

Fig. 1.3 Three legal fields for responsible robots

Once such possible candidates for a hard case in the laws of robots are grasped, we have to augment the intricacy of the model: "Who pays?" often means different things in such fields as criminal law, contracts, and torts. The level of autonomy that at times is sufficient to produce relevant effects in the field of contractual obligations (that is, "I-2," "SL-2," and "UD-2"), arguably is insufficient to bring robots before judges and have them declared guilty in criminal courts (*e.g.*, "SL-1"). Likewise, when considering robots as a source of responsibility for other agents in the system ("I-3," "SL-3," and "UD-3"), attention should be drawn to the different ways we say an "agent pays its debt." In criminal law, think of the different reasons underpinning the legitimacy of inflicting punishment, *e.g.*, the theory of retribution, or of special and general prevention. In civil (as opposed to criminal) law, reflect on obligations imposed by the government that can even overrule clauses and conditions of responsibility established by the parties to a contract. In tort law, individuals are held responsible for unjust damages inflicted upon third parties, that is, harm provoked to other agents in the system. This field-sensitivity suggests refining the focus of the model by grasping the specific features of each field of the law. This stricter perspective is illustrated with a new scheme in Fig. 1.3:

By increasing the resolution of this model, new (classes of) legal issues follow as a result. Chap. 3 below explores the popular debate on robotics technology and criminal law, averting Sci-Fi scenarios, *e.g.*, criminally accountable robots. After examining matters of legal responsibility and agenthood (Chap. 2), the aim is to show that robots are affecting basic tenets of the law in two different ways. First, these machines are inducing some problems that are specific to criminal law, mostly to do with clauses of immunity. Besides the immunity of military and political authorities for the use of robots in battle, we have to determine whether the behaviour of robots falls within the loopholes of the system, necessitating the intervention of lawmakers at both national and international levels, as they did in the early 1990s when establishing a new class of computer crimes. Then, a second class of legal issues concerns how the growing autonomy of robots affects key notions of the system, such as reasonability, predictability, or foreseeability, on which an individual's fault depends. Certain scholars have suggested a failure of causation, since it would be difficult to predict what types of harm may supervene (Karnow 1996). This is a class of hard cases that criminal lawyers share with experts in tort law and contracts: for example, think of clauses and conditions between private persons often crucial in determining the party who is liable for robots involved in criminal enterprises. It should be stressed that in 2010 some criminals used tiny robotic helicopters in a jewellery heist.[4] After matters of reasonable foreseeability in criminal law, such a class of hard cases has to be further examined in the fields of contracts and torts.

The starting point of Chap. 4 is the 2005 "World Robotics"-Report of the UN and the Economic Commission for Europe, mainly focusing on "robots of peace" such as environmental robots, surgical robots and edutainment robots. Here, responsibility and legal accountability for the design, construction and use of robots, are framed as a matter of risk and predictability in contractual obligations. In addition to artificial doctors and cognitive automata such as commercial software-agents, some riskier applications, *e.g.*, ZI agents and unmanned ground vehicles (UGVs), stand for a further set of hard cases. Besides a new type of legal agent for contracts (*i.e.*, "I-2," "SL-2," and "UD-2"), the ability of robots to produce, through their own intentional acts, rights and obligations on behalf of humans, entails the risk that individuals can be financially ruined by their robots' activities. Some reckon that "the best method of accident control may be to cut back on the scale of the activity" through strict liability policies (Posner 1973: 180). Yet, it is feasible to avert legislation that makes individuals think twice before using or producing robots at all: consider new models of insurance and legal accountability for such machines, *e.g.*, the "digital peculium" of robots. Contrary to traditional forms of distributing responsibility and risk, "only robots shall pay" could, at times, be a sound approach to the contract problem (Chopra and White 2011).

Chapter 5 looks at extra-contractual responsibility, *i.e.*, when robots damage third parties rather than their contractual counterparties. What common

---

[4]*Nature*, 22 September 2011, p. 399.

lawyers define as torts deals with obligations between private persons imposed by the government to compensate for damage done by wrongdoing. In the civil law tradition, this idea of extra-contractual responsibility can be traced back to the Ancient Roman-law status of Aquilian protection, as the form of responsibility stemming from the general idea that individuals are liable for unlawful or accidental damages caused to others due to personal fault. The new class of hard cases that the growing autonomy of robots is likely to induce, concerns how we should interpret a novel kind of liability for the behaviour of others. For the first time ever, legal systems will hold humans responsible for what an artificial state-transition system "decides" to do. Moreover, this kind of liability crucially depends on the different kinds of robots with which we are dealing: a robot nanny, a robot toy, a robot chauffeur, a robot employee, and so forth. This is one of the most innovative aspects in the field of the laws of robots, as traditional forms of responsibility for the behaviour of children, pets, or employees, have to be complemented with new strict liability policies (*e.g.*, Posner); or, alternatively, mitigated through insurance models, authentication systems, and the mechanism of allocating the burden of proof.

Chapter 6 brings us back to the law as meta-technology. From the different classes of hard cases as previously mentioned, it does not follow that the aim of the law to govern the process of technological innovation, necessarily falls short in coping with its own purpose. In light of Table 1.1 (*i.e.*, "Is", "SLs," and "UDs"), we can pinpoint cases and classes of specific legal disagreements and yet, most of the time, a relatively strong consensus on both the conditions of legitimacy for the design, construction and use of robots, and the consequences in terms of responsibility, can luckily be found. Paradoxically, this general agreement makes it easier to identify potential hard cases in the field. By distinguishing between concepts of personhood (*i.e.*, "I-1," "SL-1," and "UD-1"), traditional immunity ("I-3"), causation ("UD-3"), artificial agency in contracts ("I-2," "SL-2," and "UD-2"), and new types of responsibility in tort law ("SL-3"), we can determine which cases should be taken seriously or be given priority. For example, certain scholars reckon that the legal personality of robots does not seem necessary or even convenient in the foreseeable future (Sartor 2009). However, you can be a supporter of the front of Robotic Liberation and still admit that the regulation of new robotic crimes ("I-3") should have priority over the three "1s" of Table 1.1: I-1, SL-1, and UD-1.

The conclusion of this book summarizes how scholars address the challenges of this field as first coined by Asimov in the early 1940s: "robotics." More than seventy years later, it is remarkable how his plots foresee many of the crucial issues of today's debate: the legal personhood of robots, questions of logic on how the “laws of the law” have to be interpreted, up to the design of machines that should comprehend and process such sophisticated information as the current laws of war and rules of engagement. Between law and literature, the message of Asimov’s stories seems to be clear: since robots are here to stay, the aim of the law should be to wisely govern our mutual relationships.

Chapter 2
On Law, Philosophy and Technology

Now where are we?
Exactly at the explanation. The conflict between the various rules is ironed out by the different positronic potentials in the brain.

Isaac Asimov, Runaround

Abstract What a new generation of issues concerning robotic crimes, contracts, and torts have in common is the legal quest to define who is responsible for a robotic act or omission: when something goes wrong, "Who Pays?" Lawyers accordingly determine different levels of responsibility and agency in the field of legal robotics, by ascertaining whether such autonomous and even "intelligent" machines should be reckoned as legal persons, proper agents, or mere sources of legal responsibility in the system. Three different scenarios for a hard case in positive law concern the personhood of robots, their accountability in contracts, and new types of human responsibility for the behaviour of others. However, "Who pays?" often means different things in such fields as criminal law, contracts, and torts, *e.g.*, the level of robotic autonomy that at times is sufficient to produce relevant effects in the field of contractual obligations, arguably is insufficient to bring robots before judges and have them declared guilty in criminal courts.

Research within the philosophy of technology and the sociology of the law, suggesting that the law should regulate scientific research and technology, can be likened to the classical image of Achilles and the turtle. By reversing Zeno's paradox, the pace of the law seems too slow to catch up with the race

U. Pagallo, *The Laws of Robots: Crimes, Contracts, and Torts*, Law, Governance and Technology Series 10, DOI 10.1007/978-94-007-6564-1_2,

of science and technological innovation. Since Galileo's trial in 1633 to the current debate on neuroscience and bioethics, politicians and lawmakers believed otherwise. Even though we literally can arrest the pace of scientists, *e.g.*, Galileo, the argument proffered is that the race of technology is so determined and powerful that it cannot be deterred by legal means. In his telling research on *What Technology Wants* (2010), Kevin Kelly suggests why this is the case. He draws a directly proportional rule between features and outputs of technology: "the greater the number of exotropic traits we observe in a particular expression of technology, the greater its inevitability and its conviviality" (*op. cit.*, 270). Once we understand the laws under which humans have been using tools for over hundreds of thousands of years, unveiling an already written future appears feasible. Contrary to the laws of the law, the laws of technology allow us to find the logic of human evolution: starting with the hero of the ape-like tribe of early humans grasping how a bone could be used as a weapon, down to the orbital satellite in Kubrick's famous match cut in *2001: A Space Odyssey*.

This view on technology has induced one distinguished researcher from Carnegie Mellon, Hans Moravec (1999), to announce that intelligent robots will succeed humans and that we, as a species, will then face extinction. Likewise, Ray Kurzweil's *The Singularity is Near* (2005) sketches an imminent future where greater than human intelligence emerges through technological means. Whilst Kurzweil reckons that this singular event may happen by 2045, the complementary website is keen to inform us at http://singularity-2045.org/ that we should include nanobots, artificial intelligence and robotics among the main contributing factors to this singular event. Scholars therefore have to be prepared to address a new generation of legal cases and, more particularly, new types of crimes. In *How Just Could a Robot War Be?*, for example, Peter Asaro explores the hypothesis of challenges to national sovereignty and robot revolutions; in *Autonomous Robots and the Law*, Fernando Barrio speculates over robotic sex offences; in their 2007 Ethicomp paper on *Robot Thugs*, Carson Reynolds and Masatoshi Ishikawa dwell on machines that choose to commit and, ultimately, carry out a crime. According to these perspectives, new types of cases will arise with robots accountable for their regrettable actions, as the self-consciousness of robots could materialize Sci-Fi scenarios envisioning, for example, a robot revolution and hence, a new cyber-Spartacus. In addition, the meaning of traditional legal notions such as theft and homicide would change, since the factor giving rise to the culpability of an agent, *i.e.*, its *mens rea,* would be rooted in the artificial mind of a machine that really "wants."

However, as mentioned in the introduction, we need neither Sci-Fi scenarios nor techno-deterministic stances to determine that the information revolution is affecting the tenets of the law. In addition to transforming the approach of experts to legal information, *e.g.*, the development of fields such as AI and the law, technology has brought on new types of lawsuits, or modified existing ones. Consider new offences such as computer crimes (*e.g.*, identity theft) that would be unconceivable once deprived of the technology upon which they depend. Moreover, reflect on traditional rights such as copyright and privacy, both turned into a matter of the access, control, and protection of information in digital environments. By examining the legal challenges of robotics, we thus have to specify those concepts and principles of legal reasoning that are at stake. Then can we begin to determine whether the information revolution: i) affects such concepts and principles; ii) creates new principles and concepts; or, iii) does not concern them at all, the latter being the view of traditional legal scholars. In order to discern these different cases, this chapter is presented in four sections.

Next, issues of automation and AI technology which have been debated by philosophers of law and legal scholars for decades are assessed. Herbert Hart's approach to the study of the philosophy of law appears particularly useful in order to summarize the concepts and principles of legal reasoning that may be affected by the advancement of robotics technology.

Section 2.2 focuses on the principle of responsibility and, moreover, on notions of legal accountability and liability. This stricter level of analysis allows for a further determination of whether the research and development of robotics technology alters certain cornerstones of the law.

This viewpoint is deepened in Sect. 2.3 with the concept of agency and whether robots really "act." After defining responsibility, the notion of legal agency is refined as well, so as to classify different types of liability for the behaviour of robots.

In the last section of this chapter, the aim is to clarify why this level of abstraction as defined by notions of legal responsibility and agency is specifically fruitful. After all, this level of abstraction allows us to frame the traditional legal quest: "Who pays?"

2.1 The Philosophy of Law and Robots

Research in the philosophy of law and robots can be introduced with the work of Isaac Asimov. Over the past 70 years, that is, since *Runaround* in 1942, Asimov's novels on immobile robots, metallic robots, or humanoid robots, now included in *The Complete Robot* edition (Asimov 1995), have represented the reference point for the legal challenges of this technology.

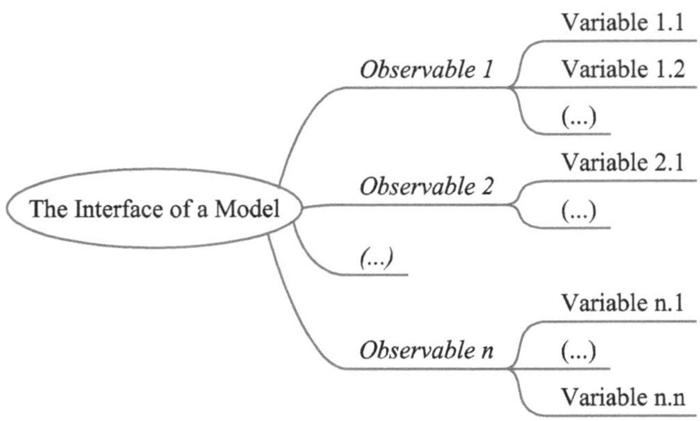

Fig. 2.1 Levels of abstraction

Moreover, they have anticipated some of the most relevant issues of today's work on the laws of robots. From a methodological viewpoint, Asimov's stories represent a fruitful level of abstraction with which the set of legal principles, concepts, and ways of reasoning we find in the laws of robots can be properly introduced. Following Luciano Floridi's remarks on *The Method of Levels of Abstraction*, the view through which one describes, examines, and argues about a given field must be chosen. The level of abstraction as the interface making an analysis of a system possible, comprises a set of features representing the observables of an analysis, the result of which provides a model for the field. The methodological approach of this book is illustrated with the first figure in this Chapter on the interface of the model, its observables and variables (Fig. 2.1):

Next, the level of abstraction concerning Asimov's work with his famous *Laws of robotics*, is illustrated in Sect. 2.1.1. The panoply of topics and legal issues derived from Asimov's stories are presented in Sect. 2.1.2 in accordance with the tripartite approach to jurisprudence that Herbert Hart gives in *The Concept of Law*. Finally, the focus in Sect. 2.1.3 is on what all the observables and variables of the analysis have in common. This stricter perspective leads to another level of abstraction, *i.e.*, the principle of responsibility as the interface of the model, discussed in Sect. 2.2.

2.1.1 The Law in Literature

Asimov conceived the three laws of robotics in his first robotic novel, *Runaround*, about a 2015 mission to a mining station abandoned 10 years earlier on Mercury. By the end of the story, two humans, namely Donovan and Powell, wonder why Speedy, the robot, is behaving so strangely. Although "perfectly adapted to a normal Mercurian environment," Donovan claims that Speedy seems "drunk." After reflecting on the reasons for such bizarre behaviour, Powell finally realizes why the robot looks inebriated: in the sober terms of computer science and engineering programming, it turned out that Law 3 drives poor Speedy back, whereas Law 2 drives him forward:

> Powell's radio voice was tense in Donovan's ear: 'Now, look, let's start with the three fundamental Rules of Robotics – the three rules are built mostly deeply into a robot's positronic brain.'
>
> 'We have: One, a robot may not injure a human being, or, through inaction, allow a human being to come to harm.
>
> Two, a robot must obey the orders given to it by human beings except where such orders would conflict with the First Law.
>
> And three, a robot must protect its own existence, as long as such protection does not conflict with the First or Second Law (Asimov, *The Complete Robot*, ed. 1982: 271–2).

Later, in *Robots and Empire* (1985), Asimov added the 'Zeroth' law:

> 0. A robot may not injure humanity or, through inaction, allow humanity to come to harm.

A story like *Runaround* offers real insight into the nature of the law once we pay attention to the different roles the law plays in Asimov's work. In addition to Sci-Fi scenarios of intelligent machines jeopardizing national sovereignty or starting revolutions, think of the ability of Asimov's robots to produce, through their intentional acts, rights and duties of their own. The empirical finding that new types of robots can develop certain sort of self-knowledge and autonomy has in fact induced certain scholars to suggest a parallel with Asimov's stories, since today's robots would similarly affect cornerstones of the law, such as notions of legal personhood, moral agency and constitutional rights. Advocates of what this author dubs the "Front of Robotic Liberation" reckon that "in principle artificial agents should be able to qualify for independent legal personality, since this is the closest legal analogue to the philosophical conception of a person" (Chopra and White 2011: 182). As soon as we admit that today's robots are "capable of a measure of empathy" and "a type of autonomy that affords intentional actions" (Hildebrandt 2010), the result is that lawyers should be ready to take Asimov's stories seriously: "The emergence of such entities will probably require us to rethink notions of consciousness, self-consciousness and moral agency" (Hildebrandt et al. 2010: 559).

A further parallel between law and literature is proposed by the problems of interpretation that drive Speedy back and forward in *Runaround*. The vagueness of language, and how the circumstances of a case can affect the ways by which we interpret the meaning of such general rules, have indeed paralyzed Speedy: The robot cannot decide whether it should "protect its own existence" (Law 3), or "obey the order given to it by human beings" (Law 2). Some, as Roger Clarke (1994), have proposed addressing the gaps of Asimov's normative system through a number of implicit laws: for example, a second section to Law 2 should be added, so that "a robot must obey orders given it by super-ordinate robots." Others stress an even stronger parallel between the law and literature: the use of literary work, such as Asimov's, can improve our understanding of the legal phenomenon, because of the narrative nature that characterizes both fields. This is the way Ronald Dworkin grasps the connection in *Law's Empire* (1986), likening the making of common law jurisprudence to a sort of chain novel. From this point of view, judges are like "a group of novelists [that] writes a novel *seriatim*; each novelist in the chain interprets the chapters he has been given in order to write a new chapter, which is then added to what the next novelist receives, and so on" (*op. cit.*, 229).

However, the law plays further roles in Asimov's work. Rather than suggesting what the nature of the law should be, *e.g.*, Dworkin's law as interpretation (1982), a further set of legal issues raised by Asimov's novels revolve around how to embed rules into the positronic brains of such machines. These are the "matter of fact engineering problems" in designing robots "with safety measures," that Asimov highlights in the Introduction to *The Complete Robot* edition (1995: 9–10). Besides matters of the legal interpretation of such terms as obey, protect, and not injure humans through their missions to Mercury, *e.g.*, Asimov's laws of robotics, there is the problem of setting up and restraining the behaviour of robots through codes. Here, the engineering problems concern one of the main issues of the most dynamic and well-funded of the fields of robotics technology in the early twenty-first century. The aim of military robotics technology, in fact, is to design machines that comprehend and process such sophisticated legal information as the current laws of war and rules of engagement. Some claim we can successfully meet such challenges: as Roland Arkin affirms in *Governing Lethal Behaviour* (2007), "I am convinced that they [robot soldiers] can perform more ethically than human soldiers are capable of." Others are less optimistic: US Navy-sponsored research admits that significant troubles persist when embedding such rules in autonomous robots, as such norms are "much more complex than Asimov's laws" (Lin et al. 2007).

The different ways by which scholars refer to Asimov's laws, *e.g.*, hermeneutics and military engineering, suggest that attention should be given to how we grasp the connection between the law and literature. In the case of Lin *et al.*, and a number of civilian and military laboratories around the world, scholars mention Asimov's laws to stress the current effort of engineers, computer scientists, experts of legal ontologies, and so on, to embed

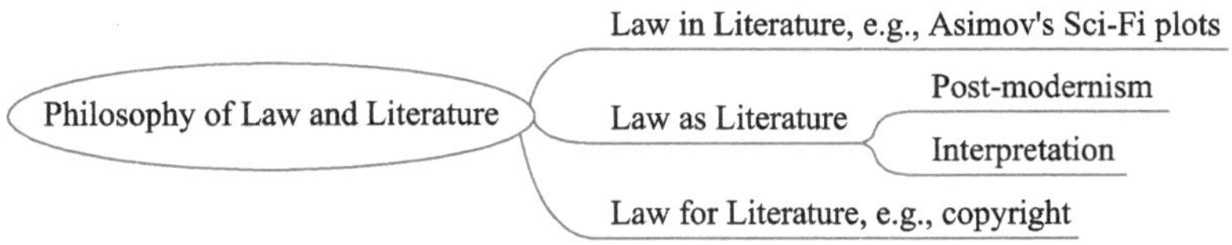

Fig. 2.2 A first model for the philosophy of law and robots

normative restraints into the on-board computers of robots. From a philosophical viewpoint, this purpose calls into question the meaning of such normative restraints. Some have proposed a parallel between Asimov's laws of robotics and the tradition of natural law, since natural law was meant to guide our actions in the same way by which the laws of robotics would direct the behaviour of robots (Comanducci 1986). Others insist on a distinction between "law as code" that may delimit or foster, but not constitute human autonomy, and "law as code" that constitutes and defines the autonomous behaviour of robots (Hildebrandt 2011). Whereas certain other scholars claim that advancements in technology would produce artificial agents capable of autonomous decisions "similar in all relevant aspects to the ones humans make" (Chopra and White 2011: 177), we should not miss the different ways by which we refer to Asimov's laws. Not taking into account a further level of analysis, such as law for literature, *e.g.*, copyright and article 27 of the 1948 Universal Declaration on Human Rights, Fig. 2.2 illustrates how scholars address law in, or as, literature:

Here, the focus is on law in literature. Rather than dwelling on post-modernist claims, *i.e.*, the narrative nature of the law as a matter of interpretation, the observables of the model concern the different classes of legal issues brought on by Asimov's stories. The interface through which how Asimov's plots anticipated (or stimulated) current research in the field of the laws of robots is illustrated with the three roads to jurisprudence as indicated by Hart in *The Concept of Law*. This does not mean we should accept any of Hart's theses, or that further differentiations are not legitimate. Rather, this is a way to show how many topics concerning the field of robotics technology can be illustrated today with a case involving an Asimov robot.

2.1.2 Sources, Concepts, and Legal Reasoning

Drawing upon Hart's approach to jurisprudence, three different kinds of legal issues can be distinguished in Asimov's robotic novels. First are the ethical issues involved in the question "What is the law?" (Hart 1961).

The well-known ability of Asimov's robots to produce, through their intentional actions, rights and obligations on behalf of humans, goes hand in hand with the claims of the Front of Robotic Liberation: the more we admit the presence of an artificial mind of a machine that affords intentional actions, the more likely it is that a new generation of ethical issues concerning the legal personhood of robots follows as a result. However, consider current research in machines ethics: the aim to build "moral machines" and teach them right from wrong seems particularly relevant in such a field as military robotics. Here, similarly to Asimov's engineers, the duty to ensure that robots are capable to abiding by principles of conduct is commonly admitted as a military necessity and for humanity, along with the aim to prevent illegal and immoral acts (*e.g.*, pillage).

The second class of problems suggested by Asimov's novels has to do with the analysis of legal concepts, such as injure and harm in the first law, command and obligation in the second law, down to the tricky notion of protection in the third law. In this context, the focus is on matters of normative hierarchy and how the legal rules are related to each other in a manner similar to that occurring with the pieces of a board game. A good illustration is offered by the aforementioned work of Roger Clarke, *Asimov's Laws of Robotics*, where he indicates various additional implicit laws with which to fill the gaps of Asimov's normative system. In particular, the First Law of Robotics should be integrated by a meta-law, which determines that "a robot may not act unless its actions are subject to the Laws of Robotics." Likewise, a new first section is proposed to be inserted in the third law, and so forth. A further illustration of this method is given in Sect. 2.2: a complex network of concepts, such as liability, accountability, burdens of proofs and clauses of immunity, complement the notions of injury and harm in Asimov's second law of robotics. On this basis, we can examine the further notion of unjust damage.

The third class of legal issues has to do with matters of interpretation and legal reasoning debated by the "law and literature movement." As Asimov's work illustrates, a proper understanding of the law is characterized by several sets of criteria for interpreting the laws of the system. Whilst robots invoked a type of literal reading in Asimov's first novel, only the extremely more sophisticated robots of the later stories began to employ complex hermeneutical techniques, such as strict or extensive interpretations of the laws, evolutionary and teleological readings of the texts, and so on. For the sake of conciseness, certain popular arguments of traditional legal hermeneutics make this point.

First, the specific property of the laws of robotics, namely their abstract and general nature, entails the difficult task of applying Asimov's laws to a

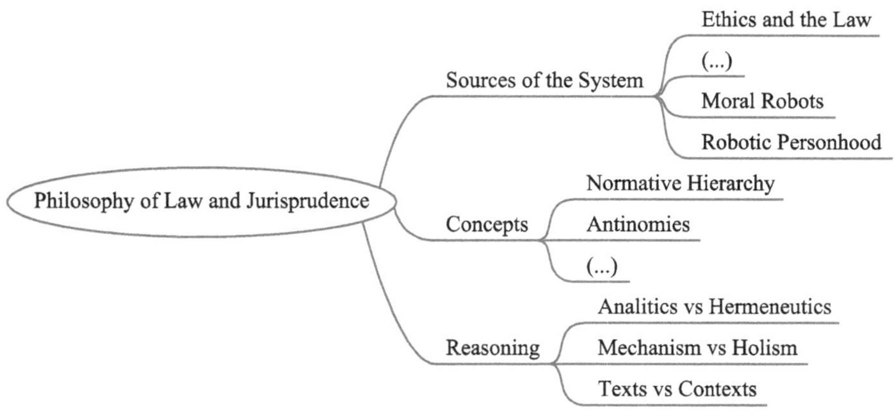

Fig. 2.3 A second model for the philosophy of law and robots

given context. Do the circumstances of the case affect how we interpret these general rules?

Second, the vagueness of ordinary language, as in the case of crucial terms like harm or order, jeopardizes the possibility of ensuring mechanical observance of the rules. Would it be feasible to develop computable models so as to comprise not only legal norms and concepts but also legal agents?

Third, adapting Hart's example of the rule that bans vehicles from a park, how about a set of criteria for grasping the meaning of a rule? Contemplate a super-market prohibiting pets: what should we think of this norm? Does this rule forbid me from bringing my favourite pet snake?

Figure 2.3 summarizes this all-encompassing view on the laws of robots that hinges on Hart's tripartite approach to jurisprudence:

In light of the legal observables in Fig. 2.3 – that is, the three ways to grasp the question "What is the law?" (Hart 1961) – let us now choose a specific problem to illustrate how the legal observables of the model are related to each other. This specific problem is suggested by the 1982 introduction to the definitive collection of robot stories, where Asimov recalls that, by the time he was in his late teens "and already a hardened science fiction reader," he used to distinguish robot stories into two classes. In contrast to the class of Robots-as-Menace, there was the class of Robots-as-Pathos concerning lovable robots that "were usually put upon by cruel human beings":

> But something odd happened as I wrote this first story [*Runaround*]. I managed to get the dim vision of a robot as neither Menace nor Pathos. I began to think of robots as industrial products built by matter-of-fact engineers. They were built with safety measures so they weren't Menaces and they were fashioned for certain jobs so that no Pathos was necessarily involved (Asimov, *The Complete Robot*, cit., 9–10).

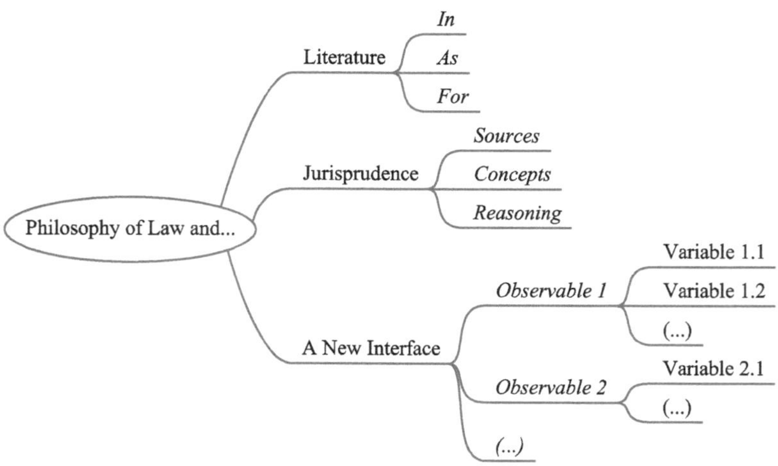

Fig. 2.4 A new interface for the philosophy of law and robots

The specific legal problem, which Asimov proposes, has to do with matters of responsibility traditionally summed up with the Latin expression, *alterum non laedere*, that is, "do not injure another." This is what typically happens in Asimov's novels, where robots either malfunction, or properly work within a set of given parameters and yet provoke harm to others. When pondering that which should legally follow in such cases, we thus have to explore how the sources of the law, concepts, and ways of legal reasoning – namely the legal observables of the model – are related to each other when a robot injures a human, or another robot. After "law and literature" as presented in Fig. 2.2, and the traditional approach to jurisprudence as given in Fig. 2.3, we now have to restrict the focus of the analysis through a new level of abstraction.

2.1.3 *The Levels of Abstraction*

Each level of abstraction, such as "law and literature" and Hart's approach to jurisprudence, can be grasped as an interface made up of a set of features, that is, the observables of the analysis. By addressing the legal challenges of robotics as a matter of responsibility, the focus is on a specific issue in the previous models: on one hand, responsibility concerns Asimov's First law of robotics and the principle that "a robot may not injure." On the other hand, attention is drawn to the hierarchical structure of the legal system, and how a complex network of concepts is at work when individuals are confronted with claims of responsibility, as portrayed in Fig. 2.4:

By changing the interface, the analysis of the new observables and variables of the model should strengthen our comprehension of the legal phenomenon, casting further light on the challenges of today's laws of robots. The aim is to restrict the focus of the previous models, to insist on the different ways the sources, concepts and reasoning of the law – that is, the legal observables of Fig. 2.3 – function when a robot provokes harm. Besides cases of responsibility concerning the design, construction and use of robots "built with safety measures" and "fashioned for certain jobs," as occur in some of Asimov's stories, what is at stake here concerns the principle established by the First Law: what are the legal observables when a robot injures? What is the set of notions at work? How are they applied in legal reasoning?

After the preliminary remarks of this section on the philosophy of law and robots, let us now explore the next level of abstraction on the principle of responsibility.

2.2 The Principle of Responsibility

Dealing with the notion of responsibility and the ancient maxim "not to injure another," a legal observable of the previous models, such as the hierarchical structure of the law, can fruitfully introduce an analysis on the role and logic of the principles of the system. As shown by Asimov's stories, there are certain fundamental norms, or superior values, that should be conceived as the principles of the system as they offer a standard for deciding what laws and rules have to be applied and how to understand them. Reflect on the content of Asimov's Second and Third Laws, in light of the principle of responsibility established by the First Law: while the second provision does not apply when it would conflict with the First Law, the application of the third rule cannot conflict with the First or Second Laws. Yet, the balance between Laws 2 and 3, governing Speedy's behaviour in *Runaround*, shows that a number of normative statements are connected to each other as principles of the system. All in all, that which paralyzes Speedy in *Runaround* is that which often ignites the legal debate. Some argue that the aim of the law should be to achieve certain goals to their maximum degree through the principles of the system (Dworkin 1985); others claim that we should distinguish between principles and values. For example, in *Facts and Norms* (1996), Jürgen Habermas affirms that principles should be deemed as normative statements having a deontological, rather than teleological meaning, because principles (such as the principle of legal responsibility) follow the logic of yes or no, or that which is for the good of all, contrary to the logic of that which is good for us, or good more or less, that characterizes values.

Admittedly, this binary logic of yes or no fits certain conditions of responsibility as mentioned in the introduction. Think of the Latin expression, *nullum crimen nulla poena sine lege*, that is, no punishment is legitimate without law: an individual's criminal responsibility is subordinated to the existence of a specific norm or statute in accordance with the principle of legality and its Anglo-Saxon counterpart, the rule of law. The logic of yes or no also fits cases of strict liability in the tort law field: here, the problem concerns whether individuals can be held responsible regardless of their fault or intentions. However, certain other cases of responsibility suggest that we should revert to the logic of good more or less. Reflect on the difference between absolute human rights (*e.g.*, protection from retrospective criminal penalties) and relative human rights (*e.g.*, privacy). In the former case, the logic of yes or no makes sense because, as previously stated, "no crime, nor punishment without law." Yet, in the case of relative human rights, lawyers do balance rights and interests, *e.g.*, individual privacy and, say, national security, according to the logic of more or less that characterizes the case law of the European Court of Human Rights. By balancing grades of responsibility, this approach typically is at work in the field of tort law as well. Consider various circumstances that set in motion a chain of events leading to a plaintiff's harm, where individual responsibility is apportioned because of contributory negligence. When multiple parties cause the plaintiff's harm, lawyers have to decide whether tortfeasor A is 40 % responsible, tortfeasor B 30 %, etc.

The reason why the role and logic of responsibility vary, hinges on the different conditions under which individuals find themselves confronted with the principle of "do not injure another." Instead of exploring the logic and role the principle has in the legal domain, attention should be drawn to what all the cases of robotics have in common, whereas individual responsibility may deal with: (i) clauses of immunity; (ii) strict liability; and (iii) responsibility depending on individual fault. In the phrasing of Floridi's method on the levels of abstraction, these are the legal observables of the model, upon which variants can be examined in connection with the previous issues concerning the hierarchy, role and logic of legal principles. Accordingly, individual responsibility is either defined *a priori*, that is: (i) by establishing it *ex ante* (strict liability rules), or (ii) excluding it at all (general irresponsibility via clauses of immunity); or (iii) individual responsibility is established *ex post*, by considering the circumstances of the case and notions such as negligence and the wrongful intentions of the agent. Back to the question of what should happen when a robot causes harm, concepts and ways of legal reasoning,

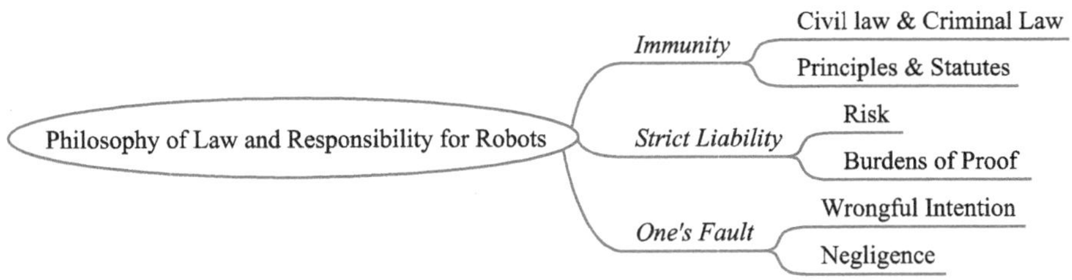

Fig. 2.5 Three conditions of responsibility for the construction and use of robots

previously illustrated with Hart's tripartite approach to jurisprudence in Fig. 2.3, can be deepened through a new interface. After the methodological remarks in Fig. 2.4, the new level of abstraction may appear as follows in Fig. 2.5:

The interface of Fig. 2.5 represents, so to speak, the statics of the system: cases of immunity, strict liability and personal fault specify the conditions by which individuals may find themselves confronted with responsibility before the law. It is thus feasible to deepen the legal observables of the previous model, namely the sources, concepts, and ways of legal reasoning as in Fig. 2.3. This is because, by dwelling on the three conditions of responsibility for the construction and use of robots as given in Fig. 2.5, we have to examine such variables as the relation between different fields of the law, the specific hierarchy between principles and statutes, as well as methods, concepts and procedures for addressing individuals' claims and rights. Therefore, the statics of the system can be illustrated through the observables of the new model: after immunity (Sect. 2.2.1), strict liability (Sect. 2.2.2), and personal fault (Sect. 2.2.3), all is ready for an analysis of the dynamics of the system, namely the legal notions of agency and agenthood (Sect. 2.3).

2.2.1 *Immunity*

The idea of legal immunity was raised in the introduction in order to address Croce's Cape Horn of legal philosophy and the difference between morals and the law. The traditional concept that "everything which is not prohibited is allowed" is summarized with the principle of legality and the corollary of the rule of law. The aim is to guarantee individual protection against arbitrary public action, so that criminal liability is imposed on the basis of specific norms in codes or statutes. This is why technological innovation continuously forces lawmakers to intervene, by adding norms for the regulation of new circumstances and new crimes. That which has happened in the field of computer crimes since the beginning of the 1990s, is likely to occur as well in the field of robotic crimes. In addition to the employment of autonomous lethal weapons in battle as mentioned in the introduction, consider a new generation of robots connected to the internet automatically collecting information in open environments, *i.e.*, out there in the real world, and bringing such environmental information to cloud servers. By replicating and spreading this data, robots could seriously impinge on current legal safeguards concerning privacy and copyright protection, trade secrets, or national security. This twofold aspect of the principle of legality, *e.g.*, immunity for cyber-thugs in the early 1990s, revolves around whether new technological applications provide loopholes within the field of criminal law.

Things are different in civil law. Think about clauses of contracts and obligations, where conditions of immunity are traditionally summed up with the Latin expression, *ad impossibilia nemo tenetur*, that is, "no one is held to that which is impossible." Here, the aim is to guarantee fair play in individual interactions and protection against the arbitrary behaviour of private individuals. Contrary to criminal law, analogy plays a crucial role in this field, as the tenet, say, of the voidability of contracts between humans could legitimately apply to artificial agents. Such a form of irresponsibility should be distinguished from cases where immunity is established *ex post*, that is, what US lawyers traditionally call "affirmative defences," in order to stress the circumstances that a defendant might raise that would excuse her liability. In addition to clauses of voidability, contemplate the annulments for mistakes in contracts, *e.g.*, mistakes relating to the substance of the subject matter of a contract, or mistakes as to the value or market price of an item. Following Giovanni Sartor's remarks in *Cognitive Automata and the Law* (2009), humans arguably would not be able to avoid the usual consequence of robots making a decisive mistake, *i.e.*, the annulment of a contract, when the human counterpart should have been aware of the mistake due to any erratic robotic behaviour.

Finally, it should be clear that lawmakers can establish in both civil and criminal law further forms of immunity by statute and what common lawyers call safe harbour-clauses. Again, the meaning of these clauses varies according to the field of the legal system. In common law, immunity of political authorities and liability of private contractors in the field of military robotics technology are defined by such norms as prescribed by the US Federal Tort Claims Act, 28 U.S.C. §§ 2401 b and 2671. Here, the Federal Tort Claims Act bars lawsuits involving discretionary law enforcement functions and different types of intentional torts. In EU law, an example is given by Article 15 of the directive 2000/31 on e-commerce: in this case, we find "no general obligation to monitor the information which [Internet Service Providers] transmit or store, nor a general obligation actively to seek facts or circumstances indicating illegal activity." At the end of the day, is it wise to adopt such clauses of immunity in all legal fields of robotics?

2.2.2 Strict Liability

The second observable of legal responsibility refers to cases where law imposes responsibility regardless of the conduct of the tortfeasor, that is, cases of no-fault responsibility or strict liability by law. Over the centuries, this has been one of the main mechanisms through which law distributes risk and responsibility. Think of individuals' liability for the behaviour of their animals and, in most legal systems, their children. Likewise, consider the responsibility of employers such as traditional publishers who, regardless of their intention or use of ordinary care, are held liable for damages caused by their employees, such as traditional media journalists and writers. These mechanisms are similarly at work in the field of dangerous activities and liability for defective products, where there is no illicit or culpable behaviour but, say, a lack of information about certain features of the product. This is the reason for the exhaustive and sometimes strange labels on products, by which manufacturers warn about risks or dangers involving improper use, *e.g.*, of a robot.

To date, strict liability regulates the design, production and use of all robotic applications that may be deemed dangerous, for example, autonomous or semi-autonomous unmanned ground vehicles. In legal terms, dangerousness hinges on whether state-of-the-art technology provides for machines capable of acting in the same way as a reasonable person in the law of torts, which is guarding against foreseeable harm. Once a robotic application has been found to not achieve such a capability, and thus, should be deemed dangerous, "it is but a short step to draw an analogy with the liability at common [and civil] law of the owner or keeper of an animal that is either known or presumed to be dangerous to mankind" (Davis 2011). As stressed by cases of apportioned liability due to contributory negligence of the plaintiff (see Sect. 2.2 above), strict liability can however be fine-tuned (or mitigated) through the allocation of the burden of proof. Once it is shown, for example, that an animal provoked harm, owners or keepers evade responsibility either when they prove that the plaintiff voluntarily assumed the risk of the injury or, in certain legal systems, when they show that a fortuitous event occurred. Analogously, in the case of strict liability for the behaviour of children, certain legal systems grant immunity when parents prove they could not prevent that harmful behaviour. The same principle applies to producers of potentially dangerous products when they show that they carefully followed the explicit regulation and detailed guidance of official legal documents.

Yet, such legal rules often fall short in coping with the advancement of technology. Whereas certain robots may behave as the reasonable person in the field of tort law, guarding against foreseeable harm, should we amend today's strict liability policies, or should we mitigate them through the allocation of the burden of proof? Is it a matter of preventing the actions of robots, *i.e.*, robots as kids, or should we prove that a fortuitous event has occurred, *i.e.*, robots as animals? Would such responsibility vary according to the different typology of robots with which we are confronted?

2.2.3 Personal Fault

The third observable of legal responsibility hinges on that which individuals voluntarily agree upon through contracts or on damages provoked by their own fault. Most of the time, responsibility is not defined *a priori*, that is, by establishing it *ex ante* (strict liability rules), or excluding it at all (general irresponsibility via clauses of immunity). Rather, liability is established *ex post*, as occurs in tort law when the reasonable person fails to guard against foreseeable harm or a person has voluntarily performed the wrongful action prohibited by the law. This kind of liability therefore is grounded on the circumstances of the case: contrary to conditions of strict liability, the burden of proof falls on the plaintiff, who has to show either the wrongful intention of her counterparty or the negligence of the tortfeasor.

This method of determining responsibility via the burden of proof can be illustrated with the da Vinci surgeon robots and a prostatectomy that a patient underwent at the Bryn Mawr hospital in Philadelphia in 2005. During the robot-assisted intervention, the machine started displaying error messages and, what is more, did not allow the human team of doctors to manually reposition its arm. After 45 min the doctors decided to undock the robot completely, they were able to manually proceed with the surgery. Still, 1 week later, the patient suffered a serious haemorrhage and, later on, erectile dysfunction and daily abdominal pains. A lawsuit against both the Da Vinci manufacturer and the hospital was brought in the Court of Common Pleas in Philadelphia. Leaving aside details of the case which are discussed below in Sect. 4.2, what matters here is that the burden of proof did not fall on the defendants but, rather, on the plaintiff. Since figures of the da Vinci robot show that such machines operate as well as, if not better than, humans, it was the patient who had to provide convincing support for his claims, *i.e.*, the fault of his counterparties.

Based on the circumstances of the case, this way of distributing responsibility and risk does not only apply to civil law, *e.g.*, contracts. Another corollary of the principle of legality and the rule of law is that fault has to be proven by public prosecutors in criminal law according to a specific norm or statute (see Sect. 2.2.1). The reversal of this method for determining responsibility via the burden of proof has to be considered an exception. Aside from cases of no-fault liability in tort law, it is only in authoritarian regimes and Kafkaesque scenarios that defendants need to prove their innocence.

2.2.4 Responsibility for a Robot

In light of the distinction between immunity, strict liability, and circumstantial fault, the level of abstraction defined by the interface of legal responsibility summarizes that which all cases concerning the design, construction and use of robots have in common. When a robot does not properly work within a given set of parameters, it is likely that there will be counterparties raising the causes of harm. Once it is shown that a robot provoked such harm, one should ask whether it concerns: (i) clauses of legal irresponsibility (*e.g.*, use of robot soldiers under laws of war); (ii) strict liability-rules (*e.g.*, dangerous UGVs); or, (iii) the circumstances of the case, *e.g.*, specific malfunctions of the da Vinci surgeon mentioned in the previous section.

However, there is a limit to this model: That which the level of abstraction does not clarify is whether the conditions of responsibility include the liability of robots, so that robots may (or should) be recognized as being legally responsible. Leaving aside scenarios of machines choosing to carry out crimes, the hypothesis of "legally-responsible robots" might be taken seriously once we reflect on the advancement of technology. For example, the ability of artificial agents to act as online traders, to buy commodities and resell them at higher prices, suggests that no Sci-Fi is needed to imagine humans transferring to robots an amount of money to be used in online transactions: when the machine does not fulfil its obligations, its creditors could directly sue the artificial agent. In addition, work on how individuals use praise and punishment in collaborative game-scenarios with computers and anthropomorphic or zoomorphic robots shows that such machines can represent a meaningful target of human censorship. Significantly, Bartneck et al. (2006) argue that by using plus and minus points as approvals and penalties for correct or wrong partner answers, "results show that praise and punishment were used the same way for computer and human partners."

All in all, it makes a lot of sense that at least certain types of robots should be held responsible for their actions in civil law, as discussed further below in Sects. 4.3 and 4.4. Moreover, some think that it is appropriate to conceive of robots as criminally accountable for their behaviour. In *A Legal Theory for Autonomous Artificial Agents* (2011), Samir Chopra and Laurence White make this point clear when they affirm that "at the risk of offending humanist sensibilities," we should yield before the fact that sooner or later, robots will be a sort of being *sui juris*, capable of "sensitivity to legal obligations" and even "possessing a moral susceptibility to punishment" that finally allows us "to forgive a computer" (*op. cit.*, 180).

To be sure, this would not be the first time legal systems hold non-humans as legally responsible for certain kinds of harm. A popular analogy casts light on how boundaries of legal responsibility have profoundly changed over centuries: some insist on the parallel between robots and animals as sources of strict liability in the field of tort law, see Sect. 2.2.2 above. Others, such as David McFarland in *Guilty Robots, Happy Dogs*, claim that we should frame our legal relationships with robots as we do with personal fault for the behaviour of animals, rather than harm provoked by tin machines or smart fridges. But how about the possibility of both robots and animals considered as responsible for their own actions? Let me clarify the parallel with a sketchy remark on history of law:

> From the ninth century to the nineteenth in Western Europe, there are over 200 well-recorded cases of trials of animals. The animals known to have been placed on trial during this period include: asses, beetles, bloodsuckers, bulls, caterpillars, chickens, cockchafers, cows, dogs, dolphins, eels, field mice, flies, goats, grasshoppers, horses, locusts, mice, moles, pigeons, pigs, rats, serpents, sheep, slugs, snails, termites, weevils, wolves, and miscellaneous vermin.
>
> Not always did the animals win their case. Some animals were severely punished, burnt at the stake; others merely singed and then strangled before the carcass was burned. Frequently the animal was buried alive. A dog in Austria was placed in prison for a year; at the end of the seventeenth century a he-goat in Russia was banished to Siberia [!]. Pigs convicted of murder were frequently imprisoned before being executed; they were held in the same prison, and under substantially the same conditions, as human criminals (William Ewald 1995, *What Was it Like to Try a Rat?*).

Needless to say, scholars today find such rites bizarre; a sort of mix between credulity and superstition. The reason hinges on how legal responsibility is connected to the behaviour of the agent and, moreover, on the type of agent with which we are dealing in terms of immunity, strict liability, or responsibility depending on fault. As stressed in the introduction, responsibility for designers, producers and users of robots calls into question whether such machines should be understood as: (i) legal persons; (ii) proper agents; or (iii) sources of responsibility for other agents in the system. Such distinctions make clear why no lawyer would prosecute a Russian he-goat today and still, it is an open question whether robots are capable of "sensitivity to legal obligations" and even of "susceptibility to punishment" (Chopra and White 2011). Although the law might discipline the behaviour of robots as it does with animals, we should be prepared to accept a new class of actions that are not purely human nor barely animal and, yet, produce multiple relevant legal effects. The next section enriches the interface of the model by exploring matters of responsibility for the behaviour of robots as a new kind of agent in the history of the law. *Pace* the Front of Robotic Liberation, such novel forms of agency do not only concern the legal personhood of robots with rights (and duties) of their own.

2.3 Agency and Accountability of Artificial Agents

After examining the statics of the model, *i.e.*, the observables of legal responsibility, this section dwells on the legal notions of agency, agenthood and personhood, that is, the dynamics of the model. Here we can take sides as to whether robotics technology: (i) affects concepts and principles of legal systems; (ii) creates new principles and concepts; or, according to a popular claim of traditional jurisprudence, (iii) does not concern them at all. First, the attention should be given to whether robots really act. Exploring the meaning of agency, and the kind of agent a robot should be perceived as, sheds light on why lawyers commonly admit that liability for harm provoked by robots should be likened to the individual accountability for the behaviour of animals, rather than cases of strict liability for dangerous products as discussed in the previous section. Some, as Michael Wooldridge and Nicholas Jennings (1995), reckon that robots, as well as any other artificial agent, enjoy such properties as autonomy, reactivity, pro-activeness and social ability to interact with other agents. Likewise, in the analysis of Stan Franklin and Art Graesser (1997), all kinds of robots are presented as reactive, autonomous, goal-oriented, mobile and temporally continuous, even though certain applications can be communicative, flexible and capable of learning and possessing a specific character. It suffices to recall the diva-bot pop star singer HRP-4C presented in the introduction.

In this context, let me emphasize the criteria pointed out by Colin Allen, Gary Varner, and Jason Zinser (2000), and further developed by Luciano Floridi and Jeff Sanders (2004), to illustrate the impact of robotics on issues of legal agenthood and, hence, on matters of responsibility before the law. Three features of robotic behaviour have to be examined so as to grasp why lawyers liken robots to animals rather than products and things:

- First, robots are interactive as they do perceive their environment and respond to stimuli by changing the values of their own properties or inner states;
- Second, robots are autonomous, because they modify their inner states or properties without external stimuli, thereby exerting control over their actions without any direct intervention of humans; and
- Third, robots are adaptable, for they can improve the rules through which their own properties or inner states change.

On this basis, the analysis of the principle of legal responsibility for the behaviour of robots deals with two different kinds of problems that can be illustrated with traditional forms of responsibility for the behaviour of animals and fellow humans. To start with, notions of moral responsibility and moral accountability must be distinguished in order to understand why today's lawyers deem superstitious that which legal systems did for centuries, *e.g.*, the trial against the poor Russian he-goat mentioned in the previous section. Once we grasp the difference between responsibility and accountability in the moral field, the second problem concerns the distinction between moral agency and the ways by which the concept of legal agency is understood, that is: (i) as a legal person; (ii) as a strict agent; and (iii) as a source of responsibility for other agents in the system. By paying attention to this tripartite distinction, light can be shed on the challenges that robotic technology poses to the traditional notion of legal agency and its variants. Although intertwined, the issues of the moral and legal agency of robots can be examined separately: it is time to sail around another Cape Horn.

2.3.1 A Moral Threshold

It strikes us as bizarre to try animals for any kind of crime or damage as legal systems did for centuries. Responsibility may be conceived of as a variant of how the notion of agency has been represented throughout the times: according to the current state-of-the-art, respondents ought to be subject to the ordinary process of moral assessment in order to determine whether they are guilty by law. Although a necessary condition, it thus is not sufficient that an agent acts. Legal systems require specific psychological components, such as consciousness and intentions, as a set of preconditions for attributing liability to a party in the case of a violation of the law. From this further viewpoint, animals are not the only agents considered legally not responsible. Such a status applies to fellow humans as well: think of young children who are not held responsible for their behaviour because of their emotional and intellectual immaturity. In addition, individuals with severe psychological illnesses are not held responsible for their actions because of their incapacity to fully understand their actions. The threshold is defined by any human of reasonable intelligence and certain maturity, who thereby is treated as an agent responsible for her conduct before the law.

On the other hand, the status of the lack of legal responsibility of an agent should be distinguished from the moral evaluation of this agent as a source of good or evil, that is, in the phrasing of Floridi and Sanders (2004), its "moral accountability." In the case of animals, consider that which typically occurs in criminal and tort law, where it can be relevant to determine whether, and how much, an animal is dangerous, so as to determine whether it should be killed (as judges and administrative authorities, at times, order). In the case of robots, it is still an open question whether the first homicide, that is, a human killed by a robot, occurred in Japan in 1991 as reported by *The Economist* in 2006, or, according to the opinion of Robert Freitas in *The Legal Rights of Robotics*, as early as 1979. The distinction between moral accountability and responsibility is thus critical: Although robots lack such requisites as consciousness, moral understanding and emotions, they can represent a new meaningful target of human censorship. Once the design, sale or supply of robotics technology is deemed illegal, lawmakers can just choose among one of the following suggestions by Floridi and Sanders in *On the Morality of Artificial Agents*: "(a) monitoring and modification (*i.e.*, 'maintenance'); (b) removal to a disconnected component of Cyberspace; (c) annihilation from Cyberspace (deletion without backup)."

Accordingly, we can extend the class of morally accountable agents so as to include the artificial agency of robots and still reject the idea that they are either morally responsible or criminally accountable: "it would be ridiculous to praise or blame an artificial agent for its behaviour or charge it with a moral accusation" (Floridi and Sanders 2004: 17). By distinguishing the source of relevant moral actions from the evaluation of agents as being morally responsible for a certain behaviour, *i.e.*, the aforementioned cases of children's actions or the behaviour of animals, we can assume that defendants have to have essential psychological qualities, such as consciousness, moral understanding and free will, to be both morally and legally responsible.

Otherwise, by blurring the notions of accountability and responsibility, we are forced back to the days when criminal trials were commonly performed against animals. The reason why today's legal systems may address animals as reasonable targets of human censorship and, still, perceive the case of the Russian he-goat mentioned in the previous section as bizarre, depends on the moral threshold of this section. In light of the distinction between moral accountability and responsibility, we can finally address Daniel Dennett's question: *When HAL Kills, Who's to Blame?* (1997). In the words of Floridi and Sanders we can say "that HAL is accountable – though not responsible – if it meets the conditions defining agenthood." How does this moral threshold affect the legal field?

2.3.2 Agents Before the Law

The moral threshold of legal agenthood, as well as matters of liability and accountability, should be examined in connection with three different kinds of agency, namely:

(i) Agents as proper persons with rights (and duties) of their own;
(ii) Strict agents in the business law-field (negotiations, contracts, etc.); and
(iii) Sources of responsibility for other agents in the system (*e.g.*, tort law).

In light of this tripartite notion of legal agency, we can further determine what kinds of responsibility are at stake with the design, construction and use of robots, by imagining the type of legal agent a robot can be. Since robots act, the question of the legal agenthood of robots requires the mapping of the multiple ways in which the notion is understood in the legal domain. Let me illustrate this point with a new level of abstraction: the model in Fig. 2.6 presents three legal observables with certain variants that summarize most of today's debate on whether robots should have rights of their own (legal personhood), can establish rights and duties on behalf of humans (strict agents), and whether current strict liability policies should be mitigated (robots as a source of responsibility for other agents in the system). After conditions where agents, both natural and artificial, may find themselves confronted with legal responsibility (see Fig. 2.5 above regarding the statics of the system), the different ways by which such agents act in the legal field must be deepened: the dynamics of the system. Let us have a look at Fig. 2.6:

The first observable in Fig. 2.6 concerns the notion of legal personhood, that is, whether an agent should be reckoned as a legal person with the ability to have rights and duties of her own. The legal observable presents three

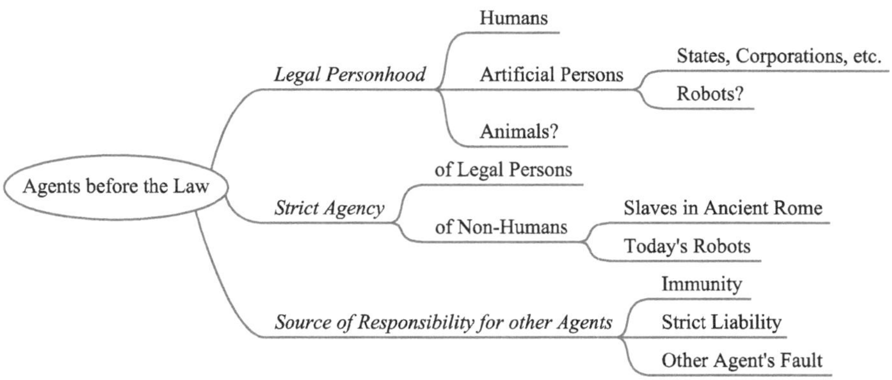

Fig. 2.6 From responsibility to legal agency and return

variants: persons can be humans, artificial persons like corporations and, according to certain scholars, animals. In the first case, Article 1 of the 1948 Universal Declaration of Human Rights summarizes today's legal framework: "All human beings are born free and equal in dignity and rights. They are endowed with reason and conscience and should act towards one another in a spirit of brotherhood." Although the responsibility of natural persons hinges on their "reason and conscience," humans cannot be deprived of their legal personhood as espoused in certain rights despite severe psychological illnesses or emotional and intellectual immaturity, *e.g.*, the 1989 UN Convention on the Rights of the Child. Therefore, a human being may have rights without responsibilities, as in the case of young children; but, since the abolition of slavery, the opposite should be deemed as simply prohibited by law. The conditions defining the legal personhood of humans also regard agents that always have rights of their own, *e.g.*, human rights, fundamental rights, etc.

The second variant of legal personhood concerns artificial persons like governments, organizations, companies or corporations. Although rights and duties of such legal persons are reducible to an aggregation of human beings as the only relevant source of their action, they are legally autonomous, in that artificial legal persons have rights and duties of their own. Consequently, lawyers debate whether artificial persons should be granted the same rights natural legal persons have, *e.g.*, the 2011 decision of the US Supreme Court on a corporation's freedom of speech under the protection of the First Amendment. Likewise, scholars focus on matters of corporate epistemology as foundational to determining their legal responsibility: on the basis of multiple accumulated actions of both humans and computers, we should ascertain what is (or should be) the information content of the corporate entity so as to determine its responsibility. Leaving aside further issues for example in corporate, tax and administrative law, that which is at stake in the case of robots is whether we should admit a new kind of artificial legal person with rights (and duties) of its own. Should lawyers resolve this issue through Turing tests? Does the legal personhood of robots depend on arguments of moral consideration? Would the legal personality of robots be unnecessary and even inconvenient in the foreseeable future?

The third variant of legal personhood has to do with today's claims on the rights of animals. This type of personhood seems closely related to that of young children, because both animals and human cubs would have rights though no specific duties or responsibilities. Interestingly, the idea that animals should be deemed as legal persons often goes hand in hand with the thesis of the legal personhood of robots, *e.g.*, Bruno Latour's *Introduction to Actor-Network-Theory* (2005) where both robots and animals are presented as new candidates for the political ecology. Whilst this perspective on the legal personification of both animals and robots draws attention to the complexity of current social systems and, correspondingly, the insufficiency of anthropological standpoints, others claim that a key difference persists. As stressed in the introduction, would robots be the hallmark of "aggressive new action centres as basic productive institutions" according to the claims of Günther Teubner in *Rights of Non-humans?*

The second observable of the model, *i.e.*, robots as proper agents, was remarkably discussed in May 2003 at the annual meeting of the American Law Institute. On that occasion, the National Conference of Commissioners on Uniform State laws proposed acknowledging the validity of contracts made by electronic agents though no action or knowledge of any human is involved. Similarly, Section 14 of the US Uniform Electronic Transactions Act proposes that "a contract may be formed by the interaction of electronic agents of the parties, even if no individual was aware of or reviewed the electronic agents' actions or the resulting terms and agreements." Besides the traditional agency of humans in the business law-field, *e.g.*, brokers, new hypotheses of agency by non-humans suggest that we should see how slaves were considered under Ancient Roman law. The ability of robots to produce, through their own acts, rights and obligations on behalf of humans brings up a new parallelism, *i.e.*, between robots and slaves, since slaves were conceived as "things" that, nevertheless, played a crucial role both in trade and commerce. May robots represent a new generation of artificial proper agents in the civil law-field? Once we accept them as such agents, would the next step be the legal personhood of these robot-traders or, contrary to the opinion of advocates of today's Front of Robotic Liberation, do they represent Teubner's "aggressive new action centres as basic productive institutions"?

The final set of observables in the model corresponds to a popular point in jurisprudence, according to which robots are neither legal persons nor proper agents. Rather, as sources of responsibility for other agents in the system, this type of robotic agency is related to the concepts of moral accountability and legal responsibility as examined in the previous section. After the patterns of traditional liability of humans for the behaviour of their animals, children or employees, robots represent a new kind of responsibility for the behaviour of others. This is indeed the first time ever legal systems will hold individuals accountable for what an artificial state-transition system decides to do. Therefore, the harm provoked by such machines should accordingly be illustrated in connection with the observables of legal responsibility mentioned above in Fig. 2.5 of Sect. 2.2. Are robots inducing new types of crimes, so that defendants will fall within the loopholes of the law and, hence, be protected by clauses of immunity and the principle of legality? When considering robots as a source of responsibility in social interaction, should we opt for forms of no-fault responsibility for the behaviour of such machines or, conversely, of negligence-related tort liability? Would it be appropriate to mitigate current regimes of strict liability or even introduce clauses of immunity, so as to prevent the risk that individuals think twice before using robots at all?

2.4 Who Pays?

As in Plato's early dialogues, a number of questions were piled up with no answers given in the previous section. The aim instead was to further refine the observables of the model rather than taking sides in today's debate on the legal personhood of robots, their agency in business law, and new forms of responsibility for the behaviour of others. Obviously, certain basic notions of the law, *i.e.*, the statics and dynamics of the system, had to be reviewed in order to determine whether (and how) robotics affects them. Let me sum up here the different steps of this analysis.

First, that which is legally at stake with the production and use of robots was introduced in light of Asimov's stories and Hart's tripartite approach to jurisprudence. Two models for the philosophy of law and robotics technology were summarized in Figs. 2.2 and 2.3 in Sect. 2.1 above.

Next, a stricter viewpoint concerning responsibility for the design, construction and use of such machines was introduced with Fig. 2.4 in Sect. 2.2, to address the fundamental issue on which party is accountable before courts. According to Fig. 2.5, the observables of legal responsibility as the statics of the model were examined, namely: (i) clauses of immunity; (ii) strict liability; and (iii) personal fault.

Finally, this outlook was deepened in connection with a moral threshold and three types of agency, that is, the dynamics of the model. Legal responsibility for the behaviour of robots varies, according to the type of agenthood with which we are dealing, namely, (i) agents as legal persons; (ii) as proper agents; or (iii) as sources of responsibility for other agents in the system. These legal observables were summarized in Fig. 2.6.

The model in Fig. 2.5 concerning responsibility as the interface of the analysis can be complemented with that of Fig. 2.6 encompassing the legal observables of agenthood. We now have nine types of legal responsibility for the behaviour of robots as Table 1.1 in the introduction above has already shown.

Let us now augment the intricacy of this model. Although the legal observables of responsibility and agency are clear, they should be considered variables of the specific field we aim to take into account. Conditions where agents find themselves confronted with responsibility, according to the different types of agenthood, vary in connection with the different tenets of criminal law, contracts and torts. Focusing on the classical question of "Who pays?," even the idea of payment represents different things according to the field of law confronted. In criminal law, individuals deserve to be punished, that is, to pay their debt to society, since criminal behaviour jeopardizes foundational elements of society, for example through murders and assaults, therefore creating social alarm. In the field of contracts, the idea of payment regards compensation to individuals affected by the harmful behaviour of a counterparty. In tort law, payment follows from obligations between private persons imposed by the state to compensate for damage provoked by wrongdoing. The different reasons why individuals ought to pay their debts to society, to contractual counterparties, or to third parties in the field of torts, have to be addressed separately, so as to properly tackle the legal challenges of robotics. Let us proceed with the analysis of the legal observables of responsibility and agency in the field of criminal law.

Chapter 3
Crimes

If he has a conscience he will suffer for his mistake.
That will be the punishment as well as the prison.

Fyodor Dostoevsky, Crime and Punishment

Abstract Robots are affecting tenets of current legal systems in a twofold way. First, robotic technology is inducing a number of critical legal loopholes, which are proper of the criminal law field, *e.g.*, the employment of autonomous robot soldiers in battle. Significantly, Christof Heyns, Special Rapporteur on extrajudicial executions, urged in his 2010 Report to the UN General Assembly that Secretary-General Ban Ki-moon convene a group of experts in order to address "the fundamental question of whether lethal force should ever be permitted to be fully automated." On the other hand, we have to determine whether the behaviour of robots falls within the loopholes of the system, necessitating the intervention of lawmakers at both national and international levels, as they did in the early 1990s when establishing a new class of computer crimes. Besides the immunity of military and political authorities for the use of robots in battle, a second class of hard cases concerns how the growing autonomy of robots affects key notions of the system, such as reasonability, predictability, or foreseeability, on which an individual's fault depends. This is the class of hard cases that criminal lawyers share with experts in tort law and contracts.

In every political system, no matter if it is the Greek *polis*, the Roman *civitas* or the modern state, individuals are punished under criminal law when their conduct jeopardizes foundational elements of society. This is true regardless of compensation to the parties harmed, because such harmful behaviour

U. Pagallo, *The Laws of Robots: Crimes, Contracts, and Torts*, Law, Governance and Technology Series 10, DOI 10.1007/978-94-007-6564-1_3,

generally speaking creates social alarm. Society's right to inflict punishment is grounded on the idea that harm affects the community as a whole, as shown by cases of murder, kidnapping or theft. Leaving aside the periods of time when animals were imprisoned, burnt at the stake, banished and the like, as seen in Sect. 2.2.4, there are a number of different reasons why punishment has been deemed legitimate throughout the centuries. Think of special and general deterrence: criminals should be punished, so as to deter them from committing further wrongs, and also to discourage other individuals from carrying out such crimes. Others reasons include the ideas of retribution, just deserts and rehabilitation: individuals may deserve to be punished either as a form of vengeance, *i.e.*, an eye for an eye, or as a way to re-educate individuals having committed an offense.

In light of the different reasons why individuals are still punished today, the aim of this chapter is to ascertain how robots may impact this framework. Consider the idea of new crimes for humans who have unjustly damaged or destroyed their robots and, *vice versa*, new types of punishment for the behaviour of robots as a meaningful target of human censorship. Moreover, we can imagine further types of offences: in the mid 1990s, the Legal Tender Project claimed that remote viewers could tele-operate a robotic system to physically alter "purportedly authentic US $ 1,000 bills" (Goldberg et al. 1996). The key point for criminal lawyers revolves around how we should interpret the behaviour of autonomous and even intelligent machines. What does it mean, for example, that robots deserve to be punished for their actions? Although retribution and just deserts can be conceived as a form of vengeance or, conversely, re-education, do such expressions make any sense in the laws of robots?

A fruitful way for tackling such issues has been pointed out by Daniel Dennett in *The Intentional Stance* (1987). This book has become a popular reference in today's debate on how to address the increasingly autonomous behaviour of artificial agents and, furthermore, matters of responsibility regarding their conduct in the legal field. As Giovanni Sartor argues in *Cognitive Automata and the Law* (2009), "the intentional stance represents usually the only possible viewpoint to explain and foresee the behaviour of complex entities that can act teleologically." Similarly, in *A Legal Theory for Autonomous Artificial Agents* (2011), Samir Chopra and Laurence White reckon that "a complex artificial agent could especially aptly be the subject of the intentional stance [as] the only coherent strategy for interacting with the agent." In this context, "intentional" stands for cognitive states such as beliefs, desires, fears or hopes. This is to be distinguished from other stances such as the physical and design stances. The approach is similar to the method of the levels of abstraction and the use of interfaces illustrated above in Sect. 2.1.3: Dennett's stances represent the ways we choose to describe, observe and argue about our subject-matter. In the case of the physical stance, for example, the aim is to explain the behaviour of an object in connection with the physical properties or conditions defined by the laws of nature. This outlook mostly concerns the inquiries of physicians and chemists when, say, they have to determine the trajectory of a missile or the reaction of a molecule. As such, the physical stance is legally at work when courts and tribunals have to ascertain the facts at trial in terms of legal causality, *e.g.*, the aforementioned trajectory of a missile launched by a robot soldier, upon which the decision of a judge may rest.

Conversely, the design stance allows us to grasp the behaviour of an object, such as a living organism or a tin machine, from the point of view of its purposes and functions. The engineering counterpart of biological evolution is given by the aim of designers to determine the form of products and processes, as well as the structure of spaces and places, to achieve a set of performances and results. Here, the physical properties of the object are insufficient or not appropriate to comprehend and predict its behaviour, as is plausible to occur when confronted with a robot soldier in battle or, alternatively, with the Japanese pop star robot singer HRP-4C on stage. In such cases, we assume that robots have been designed to undertake certain functions and will behave accordingly. If something goes wrong, it is likely that issues of reasonable foreseeability will bring us back to the opinion of experts on matters of legal causation.

However, both of these stances are inadequate to grasp the behaviour of complex agents, such as animals, humans and certain kinds of robots. Such machines are progressively capable of learning from the stimuli of their surrounding environment, gaining knowledge and skills from their own conduct, so that robots will increasingly become unpredictable not only for their users, but for their designers as well. It is often pointless to anticipate the behaviour of robots in accordance with the physical stance. All in all, even the design stance falls short when tackling the complexity of their conduct. Most of the time when dealing with robots, we should pay attention to the desires and beliefs of the agents who can act with the aim of achieving certain goals. In the wording of Dennett:

> Here is how it works: first you decide to treat the object whose behaviour is to be predicted as a rational agent; then you figure out what beliefs that agent ought to have, given its place in the world and its purpose. Then you figure out what desires it ought to have, on the same considerations, and finally you predict that this rational agent will act to further its goals in the light of its beliefs. A little practical reasoning from the chosen set of beliefs and desires will in most instances yield a decision about what the agent ought to do; that is what you predict the agent will do (Dennett, *The Intentional Stance*, 1987: 17).

Interestingly, scholars often draw opposite conclusions from this level of abstraction. Going back to *A Legal Theory for Autonomous Artificial Agents*, Chopra and White quote Dennett's illustration of the intentional stance and how it works, to ground the idea of the "independent legal personhood" of artificial agents (*op. cit.*, 12–13). *Vice versa*, in *Cognitive Automata and the Law*, Sartor refers to Dennett's stances to stress the pragmatic meaning of this perspective and concludes that "giving legal personality to SAs [software agents] does not seem at present necessary or even opportune" (*op. cit.*, 283). This disagreement depends on how we grasp notions of agency, responsibility and the very difference between criminal law and civil law in this context. Therefore, we have to ascertain how the intentions of our robots may affect the right to inflict punishment, starting with the analysis of the principles, rules and set of concepts characterizing the criminal (as opposed to the civil) law field.

The opinions of scholars who have taken the intentions of artificial agents literally by envisaging a new generation of robots choosing to commit and, ultimately, carry out a crime are examined in Sect. 3.1 below. According to certain Sci-Fi scenarios popular in the field, notions of liability necessarily change with robots becoming personally responsible for their actions and intentions. In connection with the principle of legality and the rule of law, it is likely that lawmakers should intervene with a new generation of robotics crimes much as they did in the field of computer crimes beginning in the early 1990s. Moreover, lawmakers will arguably be forced to rewrite criminal codes because of the emergence of robots that really want.

Section 3.2 revisits the *terra cognita* of today's state-of-art in legal science, which takes robots off the hook with respect to all claims of criminal liability. For the foreseeable future, these machines will be legally unaccountable before criminal courts, because they lack the set of preconditions, such as consciousness, free will and human-like intentions, for attributing liability to a party. This is not to say, however, that robots do not affect certain fundamental tenets of criminal law.

The focus in Sect. 3.3 is on the design, construction and use of robots employed on the battlefield. Robots are already affecting when and how resort to war can be justified (*ius ad bellum* or *bellum iustum*), and what can justly be done in war (*ius in bello*). This represents the first set of hard cases illustrated in this chapter: what the special rapporteur, Christof Heynes, stresses as "the fundamental question" that the UN Assembly should urgently address, namely, whether autonomous lethal force ever has to be permitted, was already mentioned in the introduction.

The civilian, rather than military, side of robotic crimes is considered in Sect. 3.4. A phenomenology in three parts illustrates how robots can partake or be used in criminal enterprises. By referring to certain traditional viewpoints on the "perpetration-by-another" liability model and the "natural-probable-consequence" liability approach, the aim is to show that robots provoke a further class of hard cases in the legal field, as such machines challenge common standpoints on what should be deemed as the natural or probable consequences of a certain behaviour. With crimes of intent, traditional forms of strict liability for the behaviour of others can successfully tackle cases of harm as induced by robots. However, with crimes of negligence, the behaviour of these machines affects basic concepts of criminal law, such as human culpability and the reasons why criminal punishment should be perceived as legitimate in such cases.

The final section of this chapter examines how responsibility may be apportioned between the designers, producers and users of increasingly smart robots and complex network-centric applications in the field of criminal law. Notions of foreseeability and legal causation are in focus so as to stress that such issues reverberate in the civil law field. Section 3.5 introduces the analysis of clauses and conditions of legal contracts crucial to determining "who pays" in the criminal law field.

3.1 Sci-Fi Scenarios

The level of abstraction defined by Dennett's intentional stance can be interpreted in two dichotomous ways. Some take the intentions of robots literally, as if such machines were capable of realizing or wishing what they are saying or doing. Others adopt the intentional stance as a fruitful way to describe and observe, from a legal viewpoint, human interaction with certain types of robots. Consider cases where we should be allowed to expect that a machine really means what it declares when making a contractual offer. For some, this contractual scenario as to the intentions of robots makes sense as it deepens our understanding, for example, of the good faith of humans, rather than robots' ability to understand what they are doing (Sartor 2009). Others think that certain machines really can grasp the legal terms of their behaviour and, moreover, humans could blame such robots when they do not keep their own word or when they commit some kind of offense (Hall 2007; Chopra and White 2011; etc.). As Gabriel Hallevy affirms in *Unmanned Vehicles* (2011), "when a [robot] establishes all elements of a specific offence, both factual and mental, there is no reason to prevent imposition of criminal liability upon it for that offence."

This viewpoint of robots having real intentions can be summed up in the jargon of criminal lawyers, by saying that the mental, rather than the factual, elements of a specific offense are at stake. As previously mentioned, robots have increasingly become involved in criminal enterprises over the past two decades (*e.g.*, the claims of the Legal Tender Project in 1996). Yet, we should distinguish between a sort of weak hypothetical stance and a strong ontological approach when examining the criminal behaviour of these machines. In the former case, robots can be conceived as if they had human *mens rea*, because this Sci-Fi scenario offers a fruitful perspective with which jurists can cast further light on certain tenets of the law. The weak hypothetical stance was at work in the previous Sects. 2.1.1 and 2, where the focus was on some of Asimov's robotic novels, *i.e.*, the law in literature. In *How Just Could a Robot War Be?*, Peter Asaro similarly dwells on Ĉapek's *R.U.R*'s robot revolution, admitting that it may "seem like a fanciful bit of science fiction." Still, "we can ask serious questions about the moral status of such revolutions according to just war theory. Let us imagine a situation in which a nation is taken over by robots – a sort of revolution or civil war. Would a third party nation have a just cause for interceding to prevent this?" (Asaro 2008: 6).

Conversely, the strong ontological stance claims that the advancement of robotics technology will produce artificial agents capable of autonomous decisions that are "similar in all relevant aspects to the ones humans make" (Chopra and White 2011: 177). As Storrs Hall holds in *Beyond AI* (2007), we should accept the idea of a robot that "will act like a moral agent in many ways," insofar as it would be "conscious to the extent that it summarizes its actions in a unitary narrative, and … has free will, to the extent that it weighs its future acts using a model informed by the narrative; in particular, its behaviour will be influenced by reward and punishment" (*op. cit.*, 348). The objection that, contrary to humans, robots are "just a programmed machine" seems flawed, since "too many similarities can be drawn between the combination of our biological design and social conditioning, and the programming of agents for us to take comfort in the proclamation we are not programmed while artificial agents unequivocally are" (Chopra and White 2011: 176).

In light of the weak hypothetical vs. strong ontological stances, the difference between notions of moral accountability and responsibility must be emphasized as seen above in Sect. 2.3.1. Praise and punishment can indeed be used in collaborative game-scenarios with computers and anthropomorphic or zoomorphic robots, so that plus and minus points can correct and improve the behaviour of both humans and machines. However, according to current state-of-art in both technology and legal philosophy, it would be meaningless to argue the criminal intentions of a robot to a court. These machines are not held responsible for their actions, because there is no such a thing as a robotic *mens rea*. Robots lack the prerequisites of criminal accountability, such as self-consciousness, free will and moral autonomy, so that it is difficult to imagine a court convicting a robot for its evil conduct. Although such machines can represent a meaningful target of human censorship, and be subject to the punitive sanctions of the law, the legitimacy of inflicting punishment in modern criminal law hardly fits today's autonomous machines. Back to theories of retribution, and special or general deterrence, what would it mean that a robot should pay its debt to society? Can we correct the moral character of an autonomous machine so that it fully understands why it ought not to repeat the evil action? What would be the point in punishing a robot so as to dissuade human beings, or other robots, from committing similar wrongs?

Moreover, for the sake of the argument, let us concede that a novel generation of robots endowed with human-like properties such as free will, autonomy or moral sense materializes. In such a case, lawyers should be ready to take seriously a whole set of new offences such as robot revolutions, rebellions, robberies and so forth. Once we accept that the culpability of the agent, *i.e.*, its *mens rea*, would be rooted in the artificial mind of a machine capable of a measure of empathy, or a type of autonomy, affording intentional actions, it is more than likely that the meaning of traditional notions such as stealing, rioting or killing will change. Still, what the content of such legal concepts will come to be seems a matter better assigned to the imagination of science fiction writers rather than the analysis of legal experts. Would an AI lawyer be an advocate of the tradition of natural law, so that rules should be viewed as an objective imperative whose infringement implies a violation of the nature of the artificial agent? Would the lawyer *vice versa* be a kind of legal realist, so that norms depend on how robots affect the overall understanding of the world as well as the environment and human-robots relations? And what about further stances of AI lawyers who, contrary to their colleagues keen to follow the Kelsenian lesson on the pure doctrine of the law, will emphasize, say, the institutional mechanism of robotic order?

To be fair, science fictional approaches to the laws of robots do not only concern harm as provoked by the *mens rea* of these machines. Rather than dwelling on robotic intentions, some Hollywood-style approaches can indeed be productive as they illustrate how the growing autonomy of robots may induce a new set of *actus rei*, that is, the material elements of a crime. In addition, this set of robotic *actus rei* can shed further light on the basic concepts defining human *mens rea*, such as fault, negligence and reasonable foreseeability. Reflect on two scenarios: in *Robot Thugs* (2007), Reynolds and Ishikawa conceive a machine, *i.e.*, the "Robot Kleptomaniac," that plans a series of robberies from local convenience stores, with the aim to steal some batteries and recharge its own. Although such a machine is endowed with free will and self-chosen goals, we can set the Sci-Fi details of the story aside and ask, for example, whether the unlawful conduct of the Robot Kleptomaniac depended on – and is (fully or partially) justifiable on the basis of – what is mandatory for survival. Likewise, we can guess whether the design of such robotic applications should be deemed, as such, illegal. Furthermore, we can imagine that such a machine was not designed, or used, to commit offenses but the robot, nevertheless, carried them out. Although we cannot hold robots personally responsible, their criminal conduct (*i.e.*, *actus reus*) may ultimately impact the notion of human culpability (*i.e.*, *mens rea*).

On the other hand, consider Richard Epstein's novel *The Case of the Killer Robot* (1997). Here, Robbie CX30 kills Bart Matthews and still, the homicide remains a matter of human responsibility though the *actus reus* is constituted by Robbie CX30 assassinating Bart Matthews, as the fault or *mens rea* is either of the Silicon Valley programmer indicted for manslaughter or of the company, Silicon Techtronics, which promised to deliver a safe robot. Whilst it would be pointless to put poor Robbie on trial for murder, what *The Case of the Killer Robot* suggests is paying attention to how the autonomous conduct of Robbie can affect the way we conceive the criminal liability of Silicon Techtronics, of Robbie's programmer, etc. By distinguishing between the moral accountability of the robot and the criminal liability of the human, the next section focuses on matters of *actus reus* and *mens rea*, leaving Sci-Fi scenarios aside.

3.2 The States of Mind and Criminal Acts

The cases of the Robot Kleptomaniac, Robbie CX30 and many others, suggest an inspiring connection between new (forms of existing) crimes and how the behaviour of autonomous robots may affect an individual's *mens rea*. Once the distinctions are grasped between the criminal mind of humans and the criminal conduct of robots, between legal responsibility and moral accountability, between human intentions and the cognitive states of robots, this stricter perspective allows us to determine who should be held responsible when, say, Robbie kills, or the Robot Kleptomaniac carries out a series of robberies in the neighbourhood. The traditional question of "Who pays?" raises three different kinds of issues related to the criminal conduct of robots: crimes of intent, negligence, and new types of crimes.

First, we should pay attention to the design of such machines: robots, as the Robot Kleptomaniac, can be conceived of and constructed with only the aim of having a recurrent urge to steal and perhaps, to fence stolen batteries to other robots. In the words of the US Supreme Court on technological innovation, what is at stake here concerns whether such robotic applications are "capable of substantial non-infringing uses," that is, whether they are commonly used for lawful purposes. This is what the Justices in Washington D.C. have to ascertain from time to time: in *Sony v. Universal Studios*, the Court in 1984 had to establish whether a video recording technology, such as the Betamax, was "capable of commercially significant non-infringing uses," so that "the sale of copying equipment, like the sale of other articles of commerce, does not constitute contributory infringement if the product is widely used for legitimate, unobjectionable purposes." In the case of robots, the first step is thus to determine whether a machine has only the purpose of committing crimes, *i.e.*, crimes of intent. In such a case, the design stance supersedes any further evaluation of the robot's intentions, since each behaviour of the machine should be considered an *actus reus*. Think of the further case of the human who intended to commit a crime through the robot but due to malfunctions of the machine, the latter deviated from the plan and perpetrated some other kind of offense. Even in this case, humans would be responsible for each *actus reus* of the machine.

The second kind of legal issues have to do with robots produced by humans having no intent to carry out a crime, but nonetheless were negligent when constructing or using such robots. Here, the growing autonomy and even unpredictability of robots suggest that tenets of legal reasoning, such as notions of causation, apportioned liability and fault, can be strained. Reflect on the traditional viewpoint as to the culpability of reasonable persons who should guard against foreseeable harms. In the case of criminal behaviour by a robot (*actus reus*), it can be tricky to ascertain the responsibility of designers, producers and users of such machines (*mens rea*). Going back to the adventures of the Robot Kleptomaniac, should I be liable although I did not understand that the robot was secretly planning a series of robberies from convenience stores in the neighbourhood? In this latter case, should not the responsibility be that of the designers and producers of the evil robot?

Finally, the third kind of legal issues concerns new crimes committed by humans who unjustly damage or destroy their robots, this in order to preserve consistency between such machines and their owners. Admittedly, the focus here is not on new types of human responsibility for the behaviour of robots but rather, on novel forms of prosecution against humans due to their own wrongdoing. As "informational objects," robots and other types of artificial agents can properly be reckoned as moral patients deserving respect and protection (Floridi 2013). In a hypothetical situation in which humans unjustly damage or destroy their own artificial companions, we may thus envisage forms of prosecution. Back to the example of the Robot Kleptomaniac, suppose that the machine felt the urge to steal some batteries from local convenience stores because of the blameworthy conduct of the owner who let the robot run out of energy. Legal systems provide for a number of sanctions in cases of the intentional misuse of power, vandalism, etc. However, I concede a point, often stressed by the forefront of Robotic Liberation, according to which traditional forms of responsibility for crimes committed by humans against their autonomous machines may fall short in governing our mutual interaction. One solution could be to amend current legal rules so as to be able to charge humans for abuses of robots similar to those legal systems have established for cases of animal cruelty in past decades. An even stronger solution follows from the idea "that for a computer agent to qualify as a legal agent it would need legal personhood" (Hildebrandt 2011). With this, punishment should be even harsher: contrary to the previous forms of weak responsibility for crimes committed by humans against robots, the new strong responsibility thesis claims that crimes would be perpetrated upon agents having rights (and duties) of their own. In the opinion of the 2006 research commissioned by the UK Office of Science and Innovation's Horizon Scanning Centre ("HSC"), robots could one day demand the same rights of citizenship as humans.

This idea is not new: in *Legal Personhood for Artificial Intelligences* (1992), Lawrence Solum argues that "one cannot, on conceptual grounds, rule out in advance the possibility that AIs should be given the rights of constitutional personhood" (*op. cit.*, 1260). This conceptual possibility has animated today's ideas as espoused by the front of Robotic Liberation and the debate on whether robots should be conceived of as "moral persons" (Hall 2007); "legal persons" (Chopra and White 2011); with a "criminal mind" of their own (Hallevy 2011); with the same citizenship rights as humans (HSC 2007); and so forth. According to the weak hypothetical stance illustrated in the previous section, we can follow Solum's thought experiment on robots that claim rights of constitutional personhood. Moreover, it makes sense to imagine a novel generation of offences, such as robot slavery and sex crimes against poor robot dolls (Barrio 2008); yet averting claims that robots have minds and real intentions. This is why, *pace* the strong ontological stance, we dwelt on the new robotic crimes as described above, such as hard cases induced by the employment of robot soldiers, a new generation of robots that physically alter US dollars, tiny drones that are employed in jewellery heists, etc. All in all, it seems clear that matters of design (*actus reus*) and human culpability (*mens rea*) concerning the criminal field of the laws of robots, are more urgent than the current debate on new forms of (weak or even strong responsibility for) crimes committed by humans against their autonomous machines, in addition to the moral agenthood and legal personhood of robots, etc.

Therefore, at the risk of being lambasted for reactionary anthropocentrism, the forefront of Robotic Liberation should not have priority over the regulation of the new robotic crimes that are already affecting cornerstones of the law, such as conditions of immunity in the laws of war and concepts upon which individual accountability rests on crimes of negligence. The focus in Sect. 3.3 is thus on the design, construction and use of robots employed in battle. Then, the civilian, rather than military, side of robotic crimes is examined in Sect. 3.4: the aim is to sum up the challenges of this technology in the field of criminal law, according to a phenomenology distinguishing the design from the use of criminal robots. In connection with the autonomy of these machines and matters of accomplice responsibility and traditional negligence, the analysis finally in Sect. 3.5 focuses on how the behaviour of robots can impact the key notion of legal causation. The analysis of new possible crimes as committed by humans against their robots is postponed until Chap. 6.

3.3 Robots and Just Wars

Military robotics technology is one of the most dynamic and, by far, well-funded fields of robotics today. More than half of the AI research and development in robotics in the US is sponsored by the military, whilst the construction and deployment of such applications have skyrocketed over the past decade. Peter Singer's statistics in *Wired for War* (2009) make the point clear: "when US forces went into Iraq in 2003, the ground invasion force had no unmanned system. By the end of 2004 the number had risen to 150 or so. A year later it had reached 2,400. Today the overall US military inventory is more than 12,000."[1] Aside from the use of unmanned ground and underwater vehicles ("UUVs"), an article of *The Economist* illustrates this trend in the field of unmanned aerial vehicles ("UAVs") such as the MQ-9 Reaper and the MQ-18 Predator.[2] There has been a 1,200 % increase in combat air patrols by UAVs since 2005, and a tenfold increase in the frequency of drone strikes. Although the US largely leads worldwide research and development in UAVs, some 40 countries are currently developing autonomous

[1] *Scientific American*, July 2010, p. 39.

[2] "Flight of the Drones," 8 October 2011, p. 32.

weapons and other types of robot soldiers. According to a 2011–2020 forecast by the HIS Industry Research and Analysis group, the US will invest 56 % of the global R&D in UAVs, China 12 %, Israel 9 %, Russia 8 %, Pan-European research 3 %, Britain, France, and Italy 2 % each, and so forth. As a result, no Sci-Fi imagination is necessary to suspect that the massive employment of artificial soldiers will affect (and is already impacting on) a number of crucial fields, such as the laws of war, international criminal and humanitarian law as well as constitutional law.

In order to appreciate the level of this impact, let us stand on the shoulders of giants such as Aristotle, Cicero and Vitoria. That which is legally at stake with the use of robots soldiers, can be grasped in light of four different connections between war and law:

(i) War as a means to (re-)establish the law, *e.g.*, a UN Council authorization to resort to war;
(ii) War as the object of legal discipline, *e.g.*, the Geneva Conventions from 1949;
(iii) War as a source of law, *e.g.*, revolutions; and
(iv) War as the opposite of the law, *e.g.*, Thomas Hobbes' state-of-nature.

In this context, the focus is exclusively on how military robotics technology affects the causes rendering wars just and the principles of military conduct, that is cases (i) and (ii) above. Leaving aside Sci-Fi scenarios of robotic revolutions and a Hobbesian-like robotic state-of-nature, the attention is restricted to the notion of just war dating back 2,000 years. Next, today's legal framework on the laws of war ("LOW") and rules of engagement ("ROE") are summarized in Sect. 3.3.1, so as to understand what principles and norms robot soldiers may upset. More particularly, the current debate on whether robot soldiers may legitimately kill (*bellum iustum*) in connection with parameters such as: (a) the just cause of the war, (b) violence as a last resort option, (c) reasonable success, and (d) right intention of the proper authority that enters the war is addressed in Sect. 3.3.2. The right ways to behave on the battlefield (*ius belli*) in connection with principles of military conduct such as proportionate use of force, discrimination and non-combatant immunity, down to the doctrine of the "double effect," *i.e.*, military necessity that makes collateral damages legal, are examined in Sect. 3.3.3. Finally, the legal issues of designing robots that abide by LOW and ROE, which may affect the causes of *bellum iustum,* in particular, the principle of proportionality, are raised in Sect. 3.3.4.

Over the last century, lawmakers have added a third scenario: after causes, *bellum iustum*, and conditions, *ius belli*, of just wars, there are provisions for the aftermath of warfare, *ius post bellum*. However, the classical bifurcation suffices to understand whether robot soldiers change basic tenets of today's legal framework.

3.3.1 What Robots Might Change

Two thousand years of debate on the causes that make wars just were eclipsed three centuries ago in the modern Western world: just war-theories no longer made sense after the triumph of modern legal positivism and the "paradigm of Westphalia" (1648). In the classical phrasing of Thomas Hobbes in *Leviathan*, "is annexed to the sovereignty the right of making war and peace with other nations and Commonwealths; that is to say, of judging when it is for the public good, and how great forces are to be assembled, armed, and paid for that end" (Hobbes 1651, ed. 1999). By admitting that no one is set to judge the decisions of sovereign states, no room was left to ascertain the lawfulness of the causes of war, as the law is made up by a set of rules effectively established by national sovereigns. While the immunity of sovereigns finally ended with the Nuremberg trials (1945–1946), projects for a permanent International Criminal Court ("ICC") culminated with the Treaty of Rome in October 1999 and the ICC's work in The Hague from 1 July 2002 and onwards. Far from claiming that a Kantian cosmopolitan paradigm has replaced the old legal system with current international humanitarian law, it is noteworthy that only with the end of the Cold War (1989) and the first Gulf War (1991) the topic of just wars went viral again among lawyers.

In the past two decades legal scholars have in fact increasingly debated the many conditions that make a war just: whether a legitimate claim exists, whether violence can be admitted as a last resort, whether there is a probability of success and proportionality in the use of force. Matters of proper authority have also been discussed and whether a declaration of war is always necessary. Without entering the philosophical debate on just causes of wars, let me stress how robots impact on these very causes. Consider a traditional claim of just war-advocates, that is, the right to self-defence against external attacks and whether robots affect this just reaction. While it generally is admitted that individuals have a right to self-defence, *e.g.*, protecting yourself and your family against kidnappers, pacifists question whether states would eventually have such a right to protect themselves with "great forces" (Hobbes 1999). Still, Article 51 of the UN Charter claims that "nothing in the present Charter shall impair the inherent right of individual or collective self-defence if an armed attack occurs against a Member of the United Nations." Whereas, save self-defence, force may only be used if authorized by the UN Security Council, the problem is hence determining whether military robotics-technology somehow changes this inherent right and its current regulation, namely:

(i) The 1907 Hague Convention respecting the Laws and Customs of War on Land and its annex, *i.e.*, Regulations concerning the Laws and Customs of War on Land;

(ii) The four 1949 Geneva Conventions for the Amelioration of the Condition of the Wounded and Sick in Armed Forces in the Field (Convention I); for the Amelioration of the Condition of Wounded, Sick and Shipwrecked Members of Armed Forces at Sea (Convention II); relative to the Treatment of Prisoners of War (Convention III); and relative to the Protection of Civilian Persons in Time of War (Convention IV); and

(iii) The two 1977 additional Protocols relating to the Protection of Victims of both International and Non-International Armed Conflicts.

In order to examine how legal responsibility may change with the introduction of robotics technology in warfare, let us proceed with the classical distinction between robots of just war in the sense of *bellum iustum* and robots of just war in the sense of *ius belli*. Although giants such as Aristotle, Cicero and Vitoria held these two aspects to be connected, today's scholars mostly treat them as separate issues. Therefore, we will look at how robot soldiers are changing, on one hand, the causes legitimating war and, on the other hand, the behaviour admitted in warfare. The focus will then be on how causes and conditions of just war converge in connection with the principle of proportionality.

3.3.2 *Just Causes of War*

There are two reasons why scholars reckon that robots affect the causes that make wars just, *ius ad bellum* or *bellum iustum*. First, some claim that both the autonomy and unpredictability of the behaviour of AI machines make robot-wars profoundly and irremediably unethical, because no human can ultimately be held responsible "in relation to deaths caused by an autonomous weapon system." This argument, illustrated by Robert Sparrow in *Killer Robots* (2007), has obvious repercussions in the legal field, since the capacity of robots to operate in the real world without human control would impact on a very core principle of the laws of war, such as responsibility for deaths occurring in the course of the war and, moreover, the fact that wars need to be declared by a competent authority. Indeed, if robots would cause serious harm by taking their own decisions, it is but a short step to envisaging robots that may provoke accidental wars as well. According to Armin Krishnan's remarks in another *Killer Robots*-study (2009), "this would be a very tricky case legally. The only solution would be to simply withdraw all of the AW [autonomous weapons] of this particular design," so that the further employ of this kind of robot soldier could be interpreted as a war crime or a crime against humanity.

Notwithstanding the unpredictability of such scenarios, however, it seems clear that each competent authority resorting to war is already held responsible for the behaviour of its soldiers, both humans and artificial agents, regardless of their conduct or decisions. Whilst, in the civil (as opposed to the criminal) law, there are forms of strict liability for harm caused by an individual's employees, this principle applies to military criminal law as well. When robots do not work within the limits of a given set of parameters, the fault is attributed to the manufacturers of the robot; yet, when robot soldiers operate in circumstances that make their use illegal, *e.g.*, when they do not discriminate or do apply force in disproportionate ways, no lawyer doubts that the fault has to be attributed to the military commanders and political authorities under international humanitarian law ("IHL"). As Philip Alston stressed in the 2010 Report to the UN General Assembly on extrajudicial, summary or arbitrary executions, "a missile fired from a drone is no different from any other commonly used weapon, including a gun fired by a soldier or a helicopter or gunship that fires missiles. The critical legal question is the same for each weapon: whether its specific use complies with IHL." Therefore, if it is likely that the employment of robots on the battlefield will continue to increase, as robots act quicker and store more information than humans, military commanders and competent political authorities are still held strictly responsible for all the decisions of these machines.

The second reason why robot soldiers would change the causes considered to make wars just concerns how autonomous machines lower the barriers to entry into war. In the phrasing of Peter Asaro's *How Just Could a Robot War Be?* (2008), "this is the belief that these technologies will make it easier for leaders who wish to start a war to actually start one." In the wording of *The Economist*, "a president who sends someone's son or daughter into battle has to justify it publicly, as does the congress responsible for appropriations and a declaration of war. But if no one has children in danger, is it a war?" (Drones and Democracy, October 2010).

This question highlights a relevant facet of that which is changing with the development of military robotics technology: a robot-war is still a war that, nevertheless, may lower public awareness. While civilians targeted by AI attacks often consider those who send machines to fight for them as "cowards," the reasons for sending robots on the battlefield may fade away, as shown by the new generation of drones that the US CIA's civilian counsels authorize to attack almost every day. A fully-automated military mission transforms war into a fairly technical and bureaucratic operation, risk-free so to speak, so that causes of war may also be trivial, once you imagine both armies engaging no humans but only robot soldiers. With no human in danger, would it still be a war?

There are two reasons why this second class of arguments against the use of robots on the battlefield seems flawed. On one hand, it may be argued that even the hypothesis of wars among mere robot soldiers does not theoretically affect the causes that make wars just, *e.g.*, self-defence with robots vs. aggression, or the right intention of the proper authority entering into war. On the other hand, the potential lowering of the threshold of entry into war seems typical for the advent of any significant technological advance in weapons and tactics. Although technology advancements have previously given rise to the drafting of international agreements and conventions to discipline and regulate the use of, say, chemical, biological and nuclear weapons, none of the causes determining when humans may legitimately kill appear affected by the employment of robot soldiers. This is the case of self-defence and right intention mentioned above, as well as the hypotheses of reasonable success and violence as a last resort. Therefore, it should be admitted that this is the first time ever that legal systems hold political authorities and military commissioners responsible for what robots autonomously decide to do on the battlefield. However, none of the traditional causes legitimizing war would be upset by the presence of robots in warfare. Would military robotics technology otherwise impact on the conditions that make wars just?

3.3.3 Conditions of Just Wars

Ius in bello concerns principles of military conduct, such as the proportionate use of force, discriminating between soldiers and civilians, non-combatant immunity and the doctrine of the double effect. In the opinion of several scholars, that which makes robot wars unjust hinges on the technical difficulty of designing robots so as to let them distinguish between friends and foes, civilians and combatants. The failure to do this violates the principle of discrimination and immunity as required for a just war. According to the suggestions of John S. Canning in *Weaponized Unmanned Systems* (2008), a solution could be that robots target only weapons. Likewise, following the proposal of Noel Sharkey in *Grounds for Discrimination* (2008), robot soldiers could be limited to operating only in particular regions or situations. In the phrasing of Peter Asaro (2008), "we want to design military robots in a way that allows them to refuse orders that they deem to be illegal, unjust or immoral, though researchers are only beginning to think about how we might do that."

Over the past years, multiple efforts have been made in the field. Consider, for example, the work by Roland Arkin and the Mobile Robot Laboratory at the Georgia Institute of Technology. In *Governing Lethal Behaviour* (2007), Arkin dwells on "the roboticist's duty to ensure they [*i.e.*, robot soldiers] are as safe as possible to both combatant and non-combatant" in accordance with "our society's commitment to International Conventions encoded in the Laws of War" (*op. cit.*). More particularly, the aim is to enforce this duty via a design approach, so as to program robots to act conservatively and avoid human psychological problems as with "scenario fulfilment." By developing work on deontic and modal logics, BDI models, case-based reasoning and more, the goal is to embed laws of war and rules of engagement in robot soldiers: "This implies that consideration of the LOW and ROE must be undertaken from the onset of the design of an autonomous weapon system" (Arkin 2007). As in other fields, *e.g.*, privacy by design, the approach of this project is bottom-up, in other words, starting with a small set of forbidden or obligated constraints to be incrementally developed in the further steps of the project. While both LOW and ROE determine that which is absolutely forbidden and ROE defines that which is an obligatory lethal action, robots should be programmed to be able to abide by principles of conduct, such as military necessity and humanity, and to prevent illegal and immoral acts such as pillage, unnecessary suffering of humans, unlawful targeting of military objectives and so forth: "I am convinced that they can perform more ethically than human soldiers are capable of" (Arkin 2007).

The design project comprises five different steps in order to allow a robot soldier to legally engage a target:

(i) Responsibility of humans who grant use of autonomous lethal force;
(ii) Military necessity in fixing criteria for the target;
(iii) Discrimination of the target identified as a legitimate combatant;
(iv) Principle of double intention so as to define tactics for engagement, approach and stand-off distance; and
(v) Proportionality in selecting weapon-firing patterns.

Moreover, the formalization of the project can be refined with a set of additional requirements. The principle of discrimination, for example, would require robots to be able to distinguish between civilians and combatants, between friends and foes, and to direct force only against enemy military objectives. The principle of proportionality would similarly suggest that we should program ethical robots that use only lawful weapons and employ an appropriate level of force, requiring a minimization of collateral damage, according to the principle of double intention, *i.e.*, military necessity that allows collateral damage, and so on.

However, by approaching matters of military robotics technology ethically, crucial problems persist when embedding norms such as LOW and ROE in intelligent robots. In fact, the formalization of the set of rules not only has to do with "top" normative concepts such as notions of validity, obligation, prohibition and permission. These rules present highly context-dependent normative concepts, *e.g.*, proportionality and discrimination in the use of force, which exceed today's technological capabilities. Significantly, such limits have been recognized in 2008 research sponsored by the US Department of the Navy, namely, *Autonomous Military Robotics: Risks, Ethics, and Design*. In the wording of Lin, Bekey and Abney, both laws of war and rules of engagement are "much more complex than Asimov's laws, [because] the LOW and ROE leave much room for contradictory or vague imperatives, which may result in undesired and unexpected behaviour in robots."

In addition, the lawful conduct of robot soldiers involves not only vital conditions of legitimacy for *ius in bello* such as proportionality and discrimination in the use of military force. Although Arkin claims in *Governing Lethal Behaviour* (2007) that "the advent of autonomous robots on the battlefield, as with any new technology, is primarily concerned with *Jus in Bello*" (*op. cit.*), it is likely that legal issues of designing robots that abide by LOW and ROE reverberate on the causes of *bellum iustum* as well. Let us examine this aspect of the problem separately.

3.3.4 Proportionality

Scholars often address the causes and conditions for a just war as two rigidly separated fields, for example as Michael Walzer illustrates in his classic work *Just and Unjust Wars* (1977). Even an unjust war, *e.g.*, the Nazi aggression on Poland, would indeed involve actions of soldiers that may be just or unjust, whilst military conduct in a just war may violate the overriding principle of discrimination and proportionality. In order to show, however, how causes and conditions of war may interact, let us return to *Autonomous Military Robotics* (Lin et al. 2008).

In standard perspectives on just war-theory, issues of *bellum iustum* and *ius belli* are in fact considered separately. On one hand, Lin, Bekey and Abney list among the preconditions of war necessary in order for it to be deemed just the following: (i) proper authority; (ii) just cause; (iii) proportionality; (iv) the use of force as the last option; (v) reasonable success of the war and, finally (vi) the good intention of the war-declaring authority. On the other hand, when analysing the conditions rendering conduct lawful on the battlefield, they take into account (i) discrimination and non-combatant immunity; (ii) the doctrine of double intention or effect; and (iii) proportionality. So, as a necessary condition for legal *ius ad bellum*, proportionality requires that "the good achieved by war must be proportional to the evil of waging it. Therefore, it is immoral to wage a massive war to remedy a small wrong-doing (*e.g.*, the 'Soccer War' of 1969 between Honduras and El Salvador)." Conversely, as a necessary condition for legal *ius in bello*, that is, restrictions on war-fighting techniques, proportionality means that "the military ends must be proportionate to the means: no unnecessary violence is to be used in order to attain one's military goal, but only a level of force proportionate to attaining one's goal" (*op. cit.*)

In light of this distinction, *i.e.*, proportionality as a precondition for a just war ("P1") and proportionality as a principle of military conduct ("P2"), it should be clear why the classical tradition of just war-theory, exemplified by authors such as Aristotle, Cicero and Vitoria, distinguished analytically P1 from P2, though admitting a dialectical connection (Aristotle's *Politics* VII, 1324 B). A proportionate cause to go to war, P1, may indeed be ruined by a disproportionate use of violence, P2, and *vice versa*, a proportionate use of force, P2, cannot redeem a futile reason for fighting, P1. In the field of military robotics-technology, we may agree that robot soldiers do not directly impact on P1 by, say, blurring the responsibilities of humans: theoretically speaking, robot soldiers, as previous technological advances, do not alter the golden rule, P1, that "the good achieved by war must be proportional to the evil of waging it" (Lin et al. 2007). However, the introduction of technological advances in weapons and tactics may compel us to rethink the good to be achieved by war and the proportionality of the means employed to attain that end. In the case of atomic bombs, for example, a computer simulation examined what could happen in the event of a nuclear war between India and Pakistan, each of which would be hit by fifty bombs the size of the atomic bomb dropped on Hiroshima in 1945. According to a report by *Scientific American* (January 2010), the outcome would be devastating:

(i) 20 million people would be killed on both sides of the border;
(ii) 7 million metric tons of smoke would cover the world atmosphere within two weeks;
(iii) Temperatures would drop by 2.3°F and precipitation by one-tenth; and
(iv) The global agricultural trading system would halt and around a billion people worldwide, now living on marginal food supplies, would be directly threatened with starvation.

Therefore, what good can be achieved by a war using atomic bombs? What would make a nuclear attack proportionate? Would it be "an extreme circumstance of self-defence, in which the very survival of a State would be at stake," as the International Court of Justice argued in its Nuclear Weapons Advisory Opinion (1997)? Does military robotics technology alter this scenario or, as most scholars claim, the use of robot soldiers, P2, does not affect the causes of just war, P1?

All in all, I see robots as a good example of how P1 may be ruined by P2, *e.g.*, a disproportionate use of violence due to a lack of design in embedding LOW and ROE in AI military artefacts. Significantly, the US Navy-sponsored research previously mentioned admits that "it is morally unjustifiable to deploy military robots before we have any idea of their risk to non-combatants" and even states that "we may paradoxically need to use the first deaths to determine the level of risk" (Lin et al. 2007). Likewise, the research acknowledges that "whether or not robotic weaponry will soon be able to surmount the technical challenge of this moral imperative (at least as well as human soldiers) remains unknown" (*ibid*). On this basis, going back to Arkin's design project and its five steps to legally engage in wars with robots, a crucial point of today's regulatory framework has to be maintained: military commanders as well as political authorities are responsible for the behaviour of their soldiers, regardless of what an autonomous robot may "decide" to do. Wherever such machines are deployed without the necessary testing of their reliability, harm provoked by the behaviour or decisions of robot soldiers should be interpreted as a crime against humanity or a war crime under current legal provisions of both LOW and ROE.

However, it must be admitted that today's international law is silent on the set of parameters and conditions that should strictly regulate the use of robot soldiers. Remarkably, in their 2010 Reports to the General Assembly, the UN special rapporteurs, Christof Heyns and Philip Alston, stress this point: in connection with the issues of general theory of law mentioned in the previous chapter,[3] this is indeed a case where "a reasonable compromise between many conflicting interests" should be found (Hart 1961: 128). As previous international agreements have regulated technological advancements over the past decades in such fields as chemical, biological and nuclear weapons, landmines, and the like, a similar UN-sponsored agreement is urgently needed to define the conditions of legitimacy for the employment of robot soldiers. Through a detailed set of parameters, clauses and rules of engagement, an effective treaty monitoring and verification mechanisms should allow for a determination of the locus of political and military decisions that the increasing complexity of network-centric operations, and the miniaturization of lethal machines, can make very difficult to detect (Krishnan 2009).[4]

---

[3] See above in Sect. 2.1.

[4] See the special edition of the *Journal of Law, Information & Science* (21(2)), on "Laws unmanned," with the papers of Philip Alston, Tim McCormack & Meredith Hagger, Rob McLaughlin, Mary Ellen O'Connell, Noel Sharkey, Markus Wagner, and the aforementioned work of Armin Krishnan.

Still, even if a UN-sponsored agreement may determine cases where lethal force should not be fully automated, a further set of principles, concepts and ways of legal reasoning, at stake with the governance of robot soldiers, should not depend on the content of political decisions. Besides hypotheses of bans, crimes of intent, and negligence, think of such concepts as reasonable foreseeability, fault and legal causation. The behaviour of robots does not only fall within the loopholes of humanitarian law and the laws of war, insofar as their conduct may also affect the notions on which individual responsibility is traditionally grounded in the field of criminal law, that is, harmful behaviour that jeopardizes foundational elements of society. We should therefore broaden our perspective and consider possible illegal uses of robot soldiers as a class of robots partaking or being used in criminal enterprises. After all, the impact of military robotics technology on today's legal framework is an example, albeit crucial, of the more general impact robots have on fundamental tenets of criminal law.

3.4 The Phenomenology of *Picciotto Roboto*

Over the past years, an increasing number of robots have been employed in crimes, such as the machines that physically alter US dollars, tiny drones employed in jewellery heists, unmanned underwater vehicles used by Colombian drug traffickers, and so forth. What these cases suggest is an examination of the civilian as well as the military side of robotic crimes. After the adventures of the Robot Kleptomaniac in Sect. 3.2, another figure of Reynolds and Ishikawa, that is, *Picciotto Roboto*, illustrates the different ways in which we should grasp the impact of this technology in the criminal law field. This example of Reynolds and Ishikawa concerns the use of robotic security guards as the Sohgo Security Service's Guardrobo marketed since 2005 and more particularly, the case of a security robot, namely *Picciotto Roboto*, participating in a criminal enterprise such as a bank robbery: "As such, it seems that the robot is just an instrument just as the factory which produces illegal products might be. The robot in this case should not be arrested, but perhaps impounded and auctioned" (Reynolds and Ishikawa 2007: 488).

Picciotto is the Sicilian word for those who are at the bottom of the Mafia hierarchy, thereby representing the arm, rather than the mind, of a criminal enterprise. Contrary to a traditional *Picciotto*, however, the AI properties of *Picciotto Roboto* may impact on the ways lawyers traditionally grasp individual criminal accountability as an issue of reasonable foreseeability, fault or causation. Although robots are simply a means of human *mens rea*, the crimes of such *Picciotto Robotos*, at times, challenge the reasons why punishment should be legitimate. Accordingly, we have to distinguish those cases in which humans are confronted with criminal accountability, namely, cases of bans, crimes of intent, and negligence, in order to determine whether and how robots affect such cases. This perspective deepens our previous analysis on robot soldiers, since crimes committed by such machines fall within one of the types of crimes that concern either individuals that aimed to activate or send the machine so as to commit an offense, or persons that failed to guard against foreseeable harm. In this context, let us address these cases from the very beginning, that is, starting with the set of issues stemming from the design of a criminal robot.

3.4.1 Picciotto by Design

There are cases where the design stance supersedes any evaluation of the robot's intentions, that is, cases where technology is "incapable of non-infringing uses" as discussed above in Sect. 3.2. In the 2005 *Grokster* case, for example, the US Supreme Court examined whether technologies promoting the ease of infringing on copyrights such as P2P file sharing systems were to be condemned as such, so that producers of P2P software, like Grokster and Steamcast, could be sued for "inducing copyright infringement committed by their users." In connection with the figures of the Wikipedia entry, according to which "90 % of files shared on Grokster were downloaded illegally," the point of the claimants was clear: supported by the Record Industry Association of America ("RIAA") and the Motion Picture Association of America ("MPAA"), plaintiffs claimed that infringing uses of P2P technology constitute the primary aim of such systems.

In the case of robots, the first step of the analysis is thus to ascertain whether a machine is "incapable of non-infringing uses" and, in this case, what kind of crimes follow as a result. The standard approach suggests preliminarily distinguishing between facts and valid law, so the focus is on evidentiary issues. Expert technical testimony may concern forensics in criminal trials, medical expertise to determine injuries in tort law, or economic evidence establishing losses in contractual obligations. On the basis of the evidence submitted in the case, the use of a certain technology can be deemed illegal at times: although, for example, P2P applications are constitutionally sound, the US Supreme Court found evidence that both Grokster and Steamcast took affirmative steps to foster infringement by third parties. In other cases, we can *vice versa* determine that a certain technology is simply legal, *e.g.*, the US Supreme Court decision in *Sony vs. Universal City*

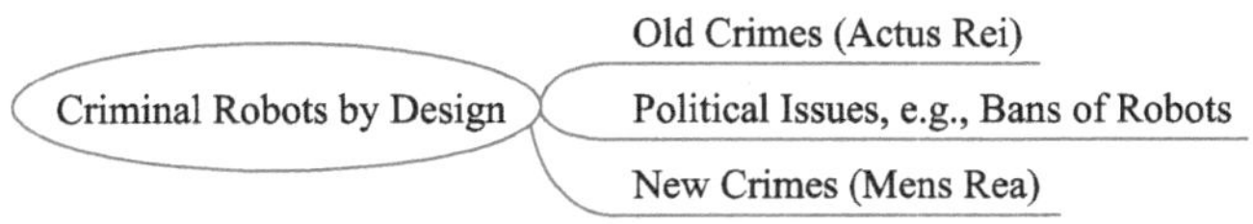

Fig. 3.1 The Phenomenology of Picciotto Roboto, step 1

Studios (1984), where the Court found evidence that Betamax and its VHR successor opened up new markets that even the plaintiffs, namely Universal, Walt Disney, Metro Goldwyn Mayer, etc., soon exploited.

Going back to the field of robotics, it is not so difficult to imagine a further set of plain cases, where there is evidence that the primary aim of a given technology is incapable of lawful uses. I already mentioned the example of robotic submarines designed and employed by Colombian drug traffickers: we can sum up this class of robots, conceived and constructed with the aim of committing some kind of offense, as *Picciotto Roboto* by design. From a criminal law viewpoint, we should distinguish between two different types of offenses (*actus reus*). First, in the case of a ban, or once ascertained that infringing uses represent the primary aim of the technology, designers, producers or users are held liable regardless of the malfunctioning of the machine or its unforeseeable and unpredictable behaviour. Every attempt to design, construct or use applications of this kind should be considered as a crime. The second type of offense concerns the additional crime perpetrated by the robot, which is conceived as if humans knowingly and wilfully committed the act. The conditions of legitimacy and responsibility for the production and use of such robots can be illustrated with Kelsen's formula "if A, then B." When the main purpose of technology is to carry out crimes (Kelsen's A), the employment of such machines is *a priori* illegal (Kelsen's B). Therefore, robots should not only be impounded and auctioned, as suggested by Reynolds and Ishikawa. Rather, it is likely that such robots should be removed to a disconnected part of cyberspace or even annihilated, as Floridi and Sanders propose in *On the Morality of Artificial Agents* (see above in Sect. 2.3.1).

Yet, we should further distinguish three hypotheses. *Picciotto Roboto* may be used either to carry out existing kinds of crimes through new robotic devices (*actus rei*), or to commit novel offences that foremost concern the versatility of human *mens rea*. However, there are cases where it could be tricky to determine what types of robots have to be banned. Figure 3.1 illustrates the new observables of the analysis:

In light of Fig. 3.1, plain from hard cases should be distinguished. Examples of plain cases are defined by the first type of legal observable of the model, such as the bank robberies of *Picciotto Roboto* by design. Here, both the conditions of legitimacy (*i.e.*, Kelsen's A), and responsibility (B) seem unproblematic, because the primary aim of a *Picciotto Roboto* by design is essentially to infringe the law. *Vice versa*, the field of military robotics technology provides examples of how evidentiary issues and matters of valid law can be far more complex. Consider the following spectrum: at one end, the MQ-18 Predator manufactured by the US based-company General Atomics; at the other end, swarms of drones that plan the mission they are going to execute by themselves. So far, responsibility for the design and construction of semi-autonomous machines such as the MQ-18 Predator depends on the technical meticulousness of the project, that is the means rather than the goals of robotic applications, on which the liability of federal contractors hinge (*e.g.*, 28 U.S.C. § 2671). Contrary to semi-autonomous machines such as the Predator, a number of autonomous systems raise issues that concern the goals, rather than the means, to be attained through such robots, *e.g.*, swarms of drones planning the mission by themselves. In the previous section, it was stressed that today's international law is silent on whether lethal force can be fully automated and what parameters and conditions should govern the use of robot soldiers. We return to this below.

The third observable of the analysis has to do with a new generation of cases concerning the *mens rea* of humans, designing specific robotic applications so as to carry out new forms of crimes. Some, as the Commissioner of the Australian Federal Police ("AFP") Mick Keelty, insist on "the potential emergence of technological crime from virtual space (online) into physical space vis-à-vis robotics."[5] Others reckon that rapid advances in robotic technology could promote "a new breed of copycat 'garden shed' robot criminals" (Sharkey et al. 2010). Here, let me stress the mania of buying and using tiny drones for civil purposes that exploded in the US in February 2012, raising threats to individual privacy, since UAVs and other types of unmanned aerial systems may collect data incessantly and, somehow, out of control. What this latter scenario suggests is a new generation of tricky cases, as the criminal intentions of humans often concern the employment of robots available at stores and shopping centres, so that the legal issue will revolve around the conditions of legitimacy for the design, production and supply of these machines, and how designers, producers and users employ such robots. Next, the focus in Sect. 3.4.2 is on this second class of crimes dependent on the use, rather than the design, of robots. By distinguishing between crimes of intent and negligence, the aim is to shed further light on whether technology should be deemed "il/legal."

[5] *Top Cop Predicts Robot Crimewave*, retrieved at http://www.futurecrimes.com/article/top-cop-predicts-robot-crimewave-2/on 31 May 2012.

3.4.2 Crimes of Intent

The first way individuals can illegally use robots concerns crimes of intent, that is, when individuals send or activate the machine in order to commit a crime. According to the current state-of-art in legal science, robots should be reckoned as innocent agents or simple instruments of an individual's *mens rea*. This is the traditional approach of criminal lawyers summed up by the "perpetration-by-another" liability model (Hallevy 2011). All in all, there are three human candidates for responsibility before a criminal court: programmers, manufacturers, and users.

First, let us dwell on the programmer of the robot with the example of Hallevy: "a programmer of AI software might design a program in order to commit offences via the AIUV [artificial intelligence unmanned vehicle]. For example: the same programmer designs software for an operating AIUV. The AIUV is intended to be placed on the road, and its software is designed to kill innocent people by running over them. The AIUV committed the homicide, but the programmer is deemed the perpetrator" (*op. cit.*) Although Hallevy's example closely resembles the hypothesis of *Picciotto Roboto* by design as seen in the previous section, we can further distinguish this scenario from the hypothesis of programmers designing a lawful AIUV operating system and, yet, using it to "run over innocent people." The difference between the two scenarios hinges on whether technological applications can widely be used for legitimate, unobjectionable purposes.

The second candidate for criminal responsibility is the manufacturer of the robot. In most legal systems, employers are held liable for any illicit action the employees engage in under their work activities under the premise of *respondeat superior*. Moreover, it is likely that both the programming and development of complex software and hardware applications, as in the field of robotics, far exceed the capabilities of a single designer. It is thus probable that lawyers will be confronted with forms of apportioned liability: still, *pace* advocates of the front of robotic liberation, the crime is a matter of human responsibility. The *actus reus* has to do with the autonomous and even intelligent behaviour of robots "running over innocent people," but the fault or *mens rea* is of the company that, say, produces killer machines and tests them in real life circumstances.

The final candidate is the user of the robot. Although the design and construction of the machine can absolutely be legal, its use may be conceived for criminal purposes. Consider the scenario of lawful civilian drones employed by the Mafia. Once again, the *actus reus* is perpetrated by the robot, but the *mens rea* is that of the user (rather than designers and manufacturers) of such machines. In the phrasing of Hallevy's *Unmanned Vehicles*, "for example, a

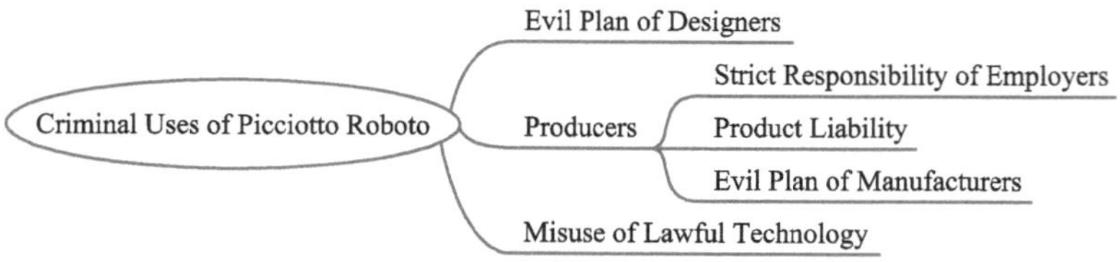

Fig. 3.2 Phenomenology of Picciotto Roboto, step 2

user purchases an AIUV, which is designed to execute any order given by its master. The specific user is identified by the AIUV as that master, and the master orders the AIUV to run over any invader of his farm. The AIUV executes the order exactly as ordered. This is not different than a person who orders his dog to attack any trespasser. The AIUV committed the assault, but the user is deemed the perpetrator" (Hallavy 2011). The second step of our phenomenology concerning the criminal uses of *Picciotto Roboto* is summarized in Fig. 3.2:

The new model of our phenomenology illustrates a set of plain cases where there is a "general agreement in judgement as to the applicability of the classifying terms" (Hart 1994: 123). The perpetration-by-another liability model allows us to properly address crimes of intent in the laws of robots, because the *mens rea* of humans makes it easy to determine who should be accountable: evil designers, faulty producers or criminal users. This is not to say, to be sure, that the miniaturization of robots or, say, the complexity of network-centric applications cannot make it very difficult to catch the human perpetrator. For example, some reckon that a new form of forensic science must be created, so that "engineers should seek ways to incorporate telltale clues into software and components to assist forensic analyses," much as "police should consider building information databases to match and trace robot crime just as they do guns and ammunition" (Sharkey et al. 2010). In light of the traditional distinction between valid law and proven facts, we will return to this below.

However, the growing autonomy of robots suggests a further set of cases where the perpetration-by-another liability model simply is useless. For instance, reflect on users intending to commit no crime through their drones but due to malfunctions of the machine, the latter does harm somehow. In such cases, lawyers have to sever the chain of responsibility and determine whether the machine properly worked within the limits of its given set of parameters or, conversely, the fault has to be attributed to the manufacturer (and designers) promising to deliver a safe machine and, yet, omitting for example certain crucial information. Moreover, there will be cases where injuries alleged by a plaintiff were actually caused by her own negligence or by her contributory negligence combined with that of an artificial agent.

From a legal viewpoint, such cases of responsibility introduce two further sets of problems.

On one hand, the focus in Sect. 3.4.3 is on cases where criminal liability hinges on negligence or lack of due care, rather than the blameworthy *mens rea* of designers, producers or users of robots. The perpetration-by-another liability model does not fit when the robot is not designed or used to carry out a specific offence, but the robot nevertheless commits it. On the other hand, Sect. 3.5 focuses on the distinction between valid law and proven facts. Whereas in the case of the *Picciotto Roboto* by design, expert technical testimony has to establish whether a robotic application is capable of non-infringing uses, a further class of legal issues concerns whether robotic applications work within a set of limits and parameters, whether the robotic behaviour can be traced back to the instructions of humans, and so forth. As advocates of the "hermeneutic circle" have stressed over the past half-century, matters of fact and proof through expert technical testimony reverberate on the way lawyers interpret the meaning of the valid law.

3.4.3 Crimes of Negligence

The final step of our phenomenology consists in cases in which criminal liability depends on negligence or lack of due care, that is, when the reasonable person fails to guard against foreseeable harms. That which Hallevy terms the "natural-probable-consequence" liability model comprises two different types of responsibility. The first scenario is closely related to the hypothesis of *Picciotto Roboto* by design, insofar as it is defined as programmers, manufacturers or users who intend to commit a crime through *Picciotto Roboto*, but the latter deviates from the plan and commits some other offence. In most legal systems, programmers, manufacturers or users of such robots would be liable for the additional crime, regardless of the unpredictability of the machine's behaviour, as it occurs with the liability model in accomplice responsibility cases. As Hallevy properly suggests, "the dangerousness of the very association or conspiracy whose aim is to commit an offence is the legal reason for more severe accountability to be imposed upon the cohorts" (*op. cit.*).

The second type of natural-probable-consequence liability is trickier since it regards humans having no intent to commit a wrong but who were negligent while designing, constructing or using a robot. For example, when machines do not work properly within the limits of a given set of parameters, the fault will be attributed to the manufacturers of such artefacts, *e.g.*, the 2008 case of the unintended movements of the Sword units employed by

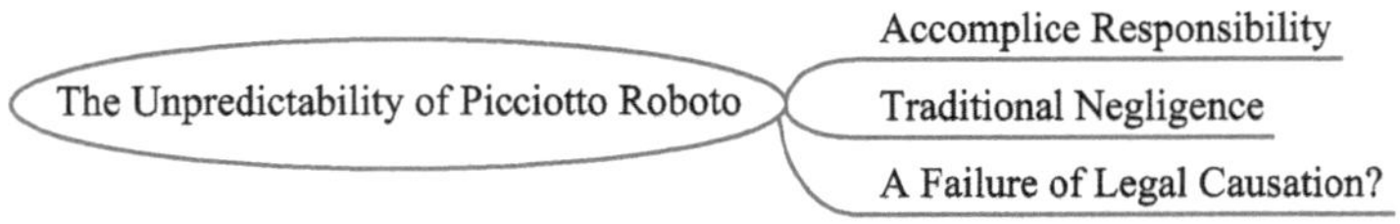

Fig. 3.3 Phenomenology of Picciotto Roboto, step 3

the US Army and claims of the producer, *i.e.*, Foster Miller, to ultimately avoid any type of liability. However, when humans reasonably fail to guard against foreseeable harms as provoked by robots, individuals are to be held responsible even when they had no intent to commit a wrong. In the view of traditional legal theory, the alleged novelty of all these cases resembles the responsibility of an owner or keeper of an animal "that is either known or presumed to be dangerous to mankind" (Davis 2011). Contrary to the criminal uses of *Picciotto Roboto*, we no longer are dealing with humans who order their robots and dogs to attack, say, any trespasser (see the previous section). Rather, as a matter of negligence, think of robots (and dogs) attacking some friends during a party in the garden of my villa. The final step of our phenomenology may look like Fig. 3.3:

By drawing an analogy between strict liability policies for damages caused by animals and human liability for the behaviour of robots, traditional legal theory acknowledges a new type of human responsibility for the behaviour of others. Since robots, like animals, do act, the result is that harms caused by robots can hardly be likened to the set of strict liability rules for dangerous activities, *e.g.*, liability for defective products and lack of information. Furthermore, since robots are machines capable of learning and adapting to changes in the environment, they are unpredictable. So, they will give rise to a new set of legal issues centred around how humans treated the machine, rather than the ways in which the machine was designed and constructed. Consider the same model of AI vehicle we are planning to buy next Christmas: By gaining knowledge or skills from their own interactions with the living beings inhabiting the surrounding environment, the same model of AI chauffeur will behave quite differently after only a few days or weeks. In the event, for example, that an unmanned ground vehicle causes harm to someone in a car accident, it is likely that we are going to have a whole new set of hard cases: How should we establish accountability when the injuries alleged by a plaintiff were caused by her own negligence? Moreover, how should we apportion liability when the injuries alleged by a plaintiff were caused by her own negligence combined with that of an artificial agent and its human master? Are strict liability rules and traditional insurance policies a sound way of addressing such scenarios? Is there an alternative scheme in order to strike a fair balance between the individual's claim to not be ruined by the decisions of their robots and the claim of a robot's counterparty to be protected when interacting with them?

All in all, it is unlikely that we will run across a single metaphor or analogy that grasps the next generation of robot-related issues of negligence in the field of criminal law. It is plausible that such liability will vary according to the different types of application being dealt with: *Picciotto Robotos*, UGVs, smart AI nannies, drones, and so forth. As against the traditional legal theory, it also seems that robots will require a normative framework of their own because, *pace* the parallel with the behaviour of animals, a failure of causation could emerge in the field. Admittedly, it is difficult to foresee what types of harm will supervene with machines responding to stimuli by changing the values of their inner states and, furthermore, improving such rules without external stimuli. Therefore, let us restrict the focus of the analysis and dwell on the ways robot may provoke or cause harm: this stricter perspective on the behaviour of robots sheds further light on both the facts and the valid law in the field of robotics.

3.5 A Failure of Causation?

Matters of legal causation are, traditionally, a nightmare for legal scholars. Lawyers have to preliminarily grasp the often extremely complex ways certain states of affairs or events come about in order to pinpoint the link between such states of affairs and the actions (or omissions) of individuals, and then determine whether those individuals should be held accountable before the courts. As previously stated in Sect. 2.2.2, there are circumstances where individuals are strictly responsible for harm or damages that concern either the behaviour of other agents, or harm provoked by inanimate objects and processes, *e.g.*, damages produced by a fire following from the collapse of a building. However, even in these cases, no-fault responsibility does not tamper with the crucial inquiry on what has actually happened and, moreover, who did what and when. Although lawyers may disagree on whether the focus should be on the substantial factor or the adequate cause in the chain of events, there must be a link between a given agency and the harm done, *e.g.*, the harm provoked by the fire that followed the collapse of a building due to, say, negligence of the constructor: "if A, then B."

Let us further address the point with the distinction between facts and valid law, that is, between natural causality and normativity. Even though terms or conditions of the law should not contradict scientific evidence on natural events, *i.e.*, the Kelsenian concept of causation, the explanatory power of science is most of the time insufficient to clarify matters of legal responsibility. The same facts can obviously be harnessed by different legal systems in divergent ways and, moreover, multiple criteria for defining the notion of causation have been developed by different legal cultures. For example, German lawyers mostly refer to the theory of the adequate event, whereas French scholars follow the theory of the strict accountability of those events. In the US, lawyers are *vice versa* divided between advocates of the but-for test and the necessary-condition test, namely, between those arguing that the action at issue in the circumstances must be necessary to the outcome, and those claiming that the action at issue instead must be a necessary part of a set of conditions sufficient for the outcome. By examining the distinction between facts and valid law, there is thus a twofold difficulty: first, lawyers have to pay attention to the fact that scientists might be debating, perhaps even divisively, on how to interpret the chain of events provoking a given state of affairs, *e.g.*, global warming. Lawyers then have the further difficulty in qualifying such an event as a necessary condition, adequate cause or sufficient reason for attributing liability to a party by a court. To make things even more complicated, some affirm that the advancement of robotic technology and, generally speaking, of autonomous artificial agents is affecting the ways lawyers think of causal connections. In the phrasing of Curtis Karnow's *Liability for Distributed Artificial Intelligence* (1996), AI agents break down the classic cause and effect analysis.

Since ancient Roman law, legal responsibility has in fact rested with the Aristotelian idea that we should take into account *id quod plerumque accidit* in the physical domain, that is, to focus on that which generally happens as the most probable outcome of a given act, fact, event or cause. By considering "an ensemble of concurrently active polymorphic intelligent agents," such as those proposed by Karnow, crucial criteria for selecting from the entire chain of events the specific condition, or the set of conditions, that best explains a given outcome, would be challenged by the unpredictable behaviour of these machines and the complexity of network-centric applications. Reflect on the example of unmanned aerial vehicles such as the Global Hawk and other machines that can operate completely by themselves. As previously stated in Sect. 3.3.2, political authorities, military commanders and public officers should be strictly responsible for the behaviour of these machines. However, we should add the responsibility of further potential defendants such as UAV operators, manufacturers, maintenance and safety contractors, contracting parties or air traffic controllers, who interact with autonomous or semi-autonomous machines, to avoid ground damage, air-to-air collisions, communication interferences, piracy, environmental concerns, down to violation of the landowner's right and claims of nuisance and trespass in tort law. The increasing capability of machines to be "independent of real time UAV-pilot control input," according to the UK Defence Standards definition of autonomous flight, severely affects the ability of lawyers to sever the chain of liability via notions of legal causation and fault. Think of how key parameters of responsibility such as foreseeable harm, or a reasonable person, change when applied to Karnow's example of "a hypothetical intelligent programming environment which handles air traffic control" such as *Alef* (Karnow 1996).

On one hand, it seems problematic to aim at determining the types of harm that may supervene with the functioning of an entire processing system such as *Alef*'s. In the phrasing of Karnow, "no judge can isolate the 'legal' causes of injury from the pervasive electronic hum in which they operate, nor separate causes from the digital universe which gives them their mutable shape and shifting sense. The result is a snarled tangle of cause and effect as impossible to sequester as the winds of the air, or the currents of the ocean" (*op. cit.*). On the other hand, the traditional idea of the reasonable person may fade away, since the duty of individuals to guard against foreseeable harms is challenged by the growing autonomy of robotic behaviour and cases where no human would be accountable for the unforeseen results of the "machine intelligence's pathology." In fact, "no human will have done anything that specifically caused harm, and thus no one should be liable for it" (Karnow 1996). Even in the simpler case of semi-autonomous aircrafts such as the MQ-1 Predator, establishing the specific responsibilities of computer programmers, software engineers, maintenance and safety contractors as well as air traffic controllers can be quite tricky.

In certain legal systems, *e.g.*, the US, robots do not break the traditional chain of causation as long as these machines are not understood as proper legal persons that can interrupt the causal link between the original agency and the harmful outcome of a chain of events. Moreover, many legal systems have addressed this crisis of the classic cause and effect analysis in the field of robots so far through strict liability policies and clauses of immunity. We already examined in Sect. 2.2.1 the condition of immunity and safe harbour clauses as one of the cases where individuals find themselves confronted with matters of legal responsibility. At the international level, conditions and clauses of immunity are established by conventions on the laws of war, humanitarian and human rights law, diplomacy, and so forth, as seen above in Sect. 3.3. At the national level, law enforcement officers are generally protected from civil rights claims insofar as their conduct does not breach constitutional norms or clearly established statutory rights. For example, the US Federal Tort Claims Act bars lawsuits involving discretionary law enforcement functions and different types of intentional torts (28 U.S.C. § 2401 b), as much as lawsuits against federal authorities premised on strict liability theories.

In addition to conditions of immunity, legal systems tackle the crisis of the classic cause and effect analysis through strict liability rules and principles of no-fault responsibility. In this Chapter we considered that which follows the *mens rea* of humans who design and construct robots to carry out crimes (steps 1 and 2 of the phenomenology of the *Picciotto Roboto*); as well as cases of accomplice responsibility for the unpredictable harm provoked by such robots (step 3 of the phenomenology). Still, Karnow's remarks on the failure of legal causation are relevant as both conditions of immunity and strict liability rules fall short in coping with a further set of issues concerning legal responsibility for the behaviour of robots. On one side, immunity policies should be conceived of as a last resort option that, moreover, in most legal systems, does not extend to state contractors (*e.g.*, 28 U.S.C. § 2671). Therefore, clauses of immunity do not prevent cases where individual responsibility might depend on fault and negligence for the design and production of such robots. On the other side, strict liability rules often go hand in hand with individual additional liability for voluntary fault or careless conduct, that is, cases where the foreseeability of the harm or the reasonableness of the human is crucial.

As a result, we should dwell on the third type of legal responsibility that is not established *ex ante* (*i.e.*, strict liability), nor excluded *a priori* (*i.e.*, immunity), since it hinges on personal fault and the circumstances of the case. Although such fault may concern either a voluntary action or the negligent behaviour of humans, it is likely that the reasonable foreseeability of the event upon which individual responsibility rests, will draw attention to the facts of the formula "if A, then B." Besides the normative viewpoint on the necessary, adequate, or but-for features of the cases, courts and tribunals, in other words, have to establish responsibility for humans, on the basis of the probabilities concerning how robots work through their on-board decision-making controllers, automatic recovery functions, communication devices, etc. In the phrasing of Caroline Foster's *Science and the Precautionary Principle* (2011), it seems apparent "that without the opportunity to come to terms with the scientific questions in a case a court or tribunal is likely to find it difficult to make findings on points such as whether a party has acted as 'necessary' or 'reasonably' in the circumstances" (*op. cit.*, 164).

Significantly, the International Court of Justice has suggested "a distinct procedure for establishing the facts" so as to improve the efficacy of international litigation (*op. cit.*, 159). Likewise, on 24 May 2005, an international *Arbitration Regarding the Iron Rhine Railway* between Belgium and the Netherlands, recommended that the parties establish a committee of independent experts to determinate the costs of reactivating the Iron Rhine

Railway: "Nor is the task of this Tribunal to investigate questions of considerable scientific complexity as to which measures will be sufficient to achieve compliance with the required levels of environmental protection" (*op. cit.*, 163). Although we do not have to buy the Tribunal's ideal of the two-stage adjudicatory procedure and the traditional difference between the facts and the valid law of the case, it is reasonable to expect that, in such cases of considerable complexity as in the field of robotics, the legal focus should preliminarily be on the scientific meaning of the machine's behaviour. Such an inquiry concerns experts in criminal law but also in AI and computer science, physics and cybernetics, neuroscience and mechanics, electronic and biology, in addition to a number of key disciplines in the humanities such as psychology.

Lawyers since the reign of ancient Roman law have luckily figured out a way to reduce such an overload of information concerning the causes of crimes by determining the set of technical questions on which individual responsibility is founded. Attention should be drawn to the clauses of the civil (as opposed to the criminal) law because, throughout the centuries, lawyers have interpreted pacts and conditions of the agreement between private individuals that define the range of responsibility of the parties, *e.g.*, the technical meticulousness of the project. Today, such contractual obligations concern a set of parameters on how a machine should work, the aims of the artificial agent, the settings of its communication and control systems, the functionality of automated recovery functions, and so on. Since the interests of the contractual parties is to restrict as much as possible the range of their responsibilities, clauses on what a reasonable safe and controllable robot may be are thus the bread and butter of those lawyers drafting contracts for the production and use of such machines. As Richard Posner argues in *Economic Analysis of Law*, we may admit that "new activities tend to be dangerous because there is little experience with coping with whatever dangers they present… The best method of accident control may be to cut back on the scale of the activity" (Posner 2007: 180). However, it is in the very interest of (the lawyers of the) designers and manufacturers of such robots not to cut back.

This self-interest of private parties pinpoints a crucial set of issues that the criminal and civil law have in common. Whilst, from a factual point of view, it can be tricky to determine who should be held accountable for the criminal behaviour of robots, we have to reflect on how lawyers grasp concepts of causation and reasonable foreseeability in the field of contracts. The interpretation of clauses and pacts between private individuals helps us to further understand the scientific meaning of the robotic behaviour as well as key notions of the field of criminal law, such as evidence and negligence.

In terms of evidence, robotics applications should be distinguished, according to the probability of events, their consequences and costs, so as to determine, or quantify, the risk for the behaviour of such machines, on which robotic crimes will often depend. Moreover, this perspective on risk and predictability in contractual obligations sheds light on further kinds of responsibility for designers, manufacturers and users of robots. If every crime committed through a robot presupposes a party who designed and built that robot, normally on the basis of a contract, the reverse is not true. Think about a panoply of civil issues, concerning clauses and pacts between private individuals, that do not involve the right to inflict punishment in criminal law. Rather, such issues have to do with the technical meticulousness of the project and the agreement on how decision-making controllers, communication devices or automatic recovery functions should work. By deepening our comprehension of the behaviour of robots, the analysis on the field of contracts strengthens our understanding on matters of causation and foreseeability that challenge today's criminal law. Although the construction and use of certain robotic applications can be considered an ultra-hazardous activity, we already have a number of machines that are reasonably safe and controllable in the field of contracts.

Chapter 4
Contracts

We scanned the skies with rainbow eyes and saw machines of every shape and size ... The sun machine is coming down, and we're gonna have a party.

David Bowie, Memory of a Free Festival

Abstract The starting point is the 2005 "World Robotics"-Report of the UN and the Economic Commission for Europe, mainly focusing on "robots of peace" such as environmental robots, surgical robots and edutainment robots. Here, responsibility and legal accountability for the design, construction, supply, and use of robots, are framed as a matter of risk and predictability in contractual obligations. In addition to artificial doctors and cognitive automata such as commercial software-agents, some riskier applications, *e.g.*, ZI agents and unmanned ground vehicles (UGVs), stand for a further set of legal hard cases. The ability of robots to produce, through their own intentional acts, rights and obligations on behalf of humans, suggests distinguishing between robots as tools of human interaction and robots as strict agents in the legal system. However, as a new form of agent in the field of contracts, the increasingly autonomous behaviour of the robot entails the risk that individuals can be financially ruined by the activities of these machines. Whereas the traditional method of accident control via strict liability policies aims to cut back on the scale of the activity, new models of insurance and legal accountability for robots, *e.g.*, the "digital peculium" of robo-traders, illustrate a sounder approach to the contract problem.

U. Pagallo, *The Laws of Robots: Crimes, Contracts, and Torts*, Law, Governance and Technology Series 10, DOI 10.1007/978-94-007-6564-1_4,

At the very beginning, they were cars. As Åke Madesäter stresses in the *Editorial* of the UN World 2005 Robotics report, "the industrial robot was first introduced in the USA in 1961 and the first applications were tested within the car industry in North America" (*op. cit.*, ix). Japanese industry began to implement this technology on a large scale in their car factories in the 1980s, acquiring strategic competitiveness by decreasing costs and increasing the quality of their products. Western car producers learned a hard lesson and followed the Japanese thinking a few years later, installing robots in their factories during the 1990s. Over the past two decades, robots have spread in both the industrial and service fields: as shown by the Report of the Economic Commission for Europe and the International Federation of Robotics (UN World Robotics 2005), we already have "machines of every shape and size," for which the Report provides an analysis on the profitability of robot investments, effects of the business cycle on such investments, the degree of concentration in different countries with prices and wages, the worldwide operational stock of different types of robots, up to the value of the world robot market in the period of 1998–2004. Admittedly, in the extremely dynamic field of robotics, such data becomes quickly out of date. However, this Report allows us to preliminarily understand the panoply of robotics applications with which we are confronted when defining clauses and conditions of contracts.

On one side, we are dealing with a class of industrial robots employed in a number of fields as different as for example, agriculture, hunting, forestry, fishing and mining. These robots are used in the manufacture of food products and beverages, textiles and leather products, wood and coke, rubber, plastic products and basic metals. They are also used when refining petroleum products and nuclear fuel, producing domestic appliances and office equipment, electrical machinery, electronic valves, tubes and other electronic components; as well as semiconductors, radio, television and communication equipment; medical precision, motor vehicles and so on. The properties of these robots can be summarized according to the ISO 8373 definition as "an automatically controlled, reprogrammable, multipurpose manipulator programmable in three or more axes, which may be either fixed in place or mobile for use in industrial automation applications" (UN 2005: 21). The programmed motions or auxiliary functions of these robots can be changed without physical alteration, that is, without the alteration of the mechanical structure or control system except for changes of programming cassettes, ROMs, etc. In connection with the axis or direction used to specify the robot motion in a linear or rotary mode, their mechanical structure suggests a further distinction between Cartesian robots, cylindrical robots, SCARA robots, articulated robots, parallel robots and so forth.

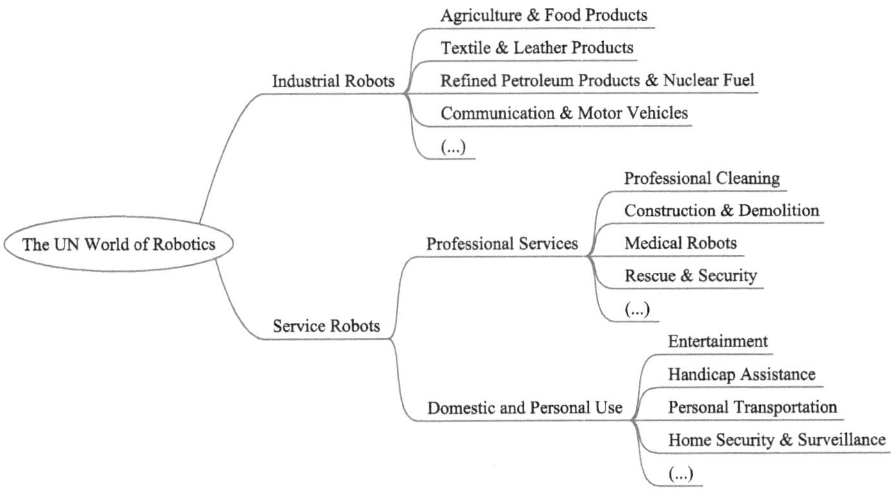

Fig. 4.1 Contractual obligations and robotics complexity

On the other side, we are also addressing a class of service robots that include professional service machines as well as domestic and personal use of robots. In the first subset, we find robots for professional cleaning, inspection systems, construction and demolition, logistics, medical robots, defence, rescue and security applications, underwater systems, mobile platforms in general use, laboratory robots, public relation robots, etc. In the second subset, there is the personal use of robots for domestic tasks such as iRobot's Roomba vacuum cleaning machines; entertainment robots such as toy robots and hobby systems; handicap assistance; personal transportation; home security and surveillance and so on. Whilst further service robots applications should be mentioned, *e.g.*, the new generation of robo-traders examined below in Sect. 4.3, such differentiations are critical in order to discern matters of responsibility and legal accountability for the design, construction, supply and use of robots, through notions of risk, safety, predictability, strict agency, delegation, and so forth, in the civil (as opposed to the criminal) law field. We can begin to chart the complexity of the field according to Fig. 4.1:

What all the robots in Fig. 4.1 have in common is a set of individual rights and obligations that on the basis of voluntary agreements between the parties to a contract concern the design, production and employment of these machines. The aim of this Chapter is to distinguish such voluntary agreements in connection with the level of risk and predictability of robotic behaviour, so as to determine whether basic concepts of contractual law, such as malfunction liability or breach of warranty, are being strained. Clauses and conditions of contracts for the construction and use of robots are differentiated next in Sect. 4.1 in light of a spectrum. At one end, there are a number of reasonable safe and controllable robots; at the other end, we find certain risky applications that can represent ultra-hazardous activities much as traditional aviation was perceived of in the 1930s.

The focus in Sect. 4.2 is on the first end of this spectrum as illustrated by the controlled settings of the operating theatres of the da Vinci surgical robots. Such machines can give rise to engineering problems that scholars routinely address as part of their research, as much as they did with previous technological innovations. On the basis of the probability of events, their consequences and costs, there is a general agreement on how lawyers should define matters of unpredictability and risk as caused by such robots, in order to ascertain individual responsibility for the design, production and use of reasonable safe machines. This class of plain (as opposed to hard) cases refers to notions of evidence, traditional negligence and cases of no-fault responsibility.

The other end of this spectrum, namely certain riskier robotic applications such as the Zero Intelligence ("ZI") agents in the business field, are the focus in Sect. 4.3. The aim is to further distinguish between robots as simple tools of human interaction and robots as proper agents in the civil law field. Although current rules bar the acceptance of the legal agency of robots in certain cases, such legal agency makes sense in that humans delegate relevant cognitive tasks to robots. These machines can send bids, accept offers, request quotes, negotiate deals and even execute contracts, so that the level of autonomy, which is insufficient to hold robots criminally accountable for their behaviour, is arguably sufficient to acknowledge new forms of artificial agency in the law of contracts.

Accordingly, Sect. 4.4 explores new forms of accountability for the behaviour of robots as well as traditional ways of distributing risk through insurance models or authentication systems. The ultimate aim is to avert legislation that makes people think twice before using or even producing robots that provide "services useful to the well-being of humans" (UN World Robotics 2005). The idea that (certain types of) robots may be held directly accountable for their own behaviour has a precedent in the ancient Roman law institution of *peculium*. In Justinian's Digest, the mechanism of *peculium* enabled slaves, deprived of personhood as the ground of individual rights, to act as estate managers, bankers or merchants. Similarly, I suggest that a sort of portfolio for robots could guarantee the rights and obligations entered into by such machines. Drawing a parallel between robots and slaves is attractive, since the aim today is the same as lawyers pursued in Ancient

Rome: individuals should not be ruined by the decisions of their robots and any contractual counterparties of robots should be protected when doing business with them.

After examining surgical robots and cognitive automata in the form of commercial software-agents, or robot-traders, Sect. 4.5 dwells on the case of unmanned vehicles and, more particularly, unmanned ground vehicles such as AI cars and chauffeurs. The reason for this is twofold. On one hand, these kinds of robotic applications allow us to deepen issues of contractual liability and both human and robotic accountability in terms of apportioned responsibility. On the other hand, AI chauffeurs suggest that we will increasingly address (or be pressed by) cases of extra-contractual responsibility, *e.g.*, robots damaging third parties rather than affecting contractual counterparties. In the event a machine fortuitously harms someone in the roundabouts, who shall pay?

4.1 Pacts, Clauses and Risk

Risk can be conceived of in three ways. First, from an evolutionary stance, we can associate the notion of risk with every adaptive attempt to reduce the complexity of the human environment. In their introduction to *Risk Analysis and Society* (2004), Timothy McDaniels and Mitchell J. Small stress that "since the beginning of human development, risks to health and well-being have led to adaptive responses that open paths for change. When Neolithic family groups shared knowledge and resources for combating hunger, thirst, climate, or outside attack, they were trying to manage risks they faced… Risk management has been a fundamental motivation for development of social and governance structures over the last 10,000 years" (*op. cit.*).

A second approach insists on the peculiar features of current modern risk societies and what therefore distinguishes them from traditional (or pre-modern) organizations as well as early modern societies. A classical text such as Ulrich Beck's 1986 *Risikogesellschaft* makes this point clear: "[W]e are eyewitnesses – as subjects and objects – of a break within modernity, which is freeing itself from the contours of the classical industry society and forging a new form – the (industrial) risk society… The argument is that, while in classical industry society the 'logic' of wealth production dominates the 'logic' of risk production, in the risk society this relationship is reversed" (1992 English edition: 9, 14).

A final approach to the notion of risk is methodological: we have to determine the level of risk through quantitative and qualitative evaluations of safety factors, risk assessment and management in terms of probabilities, engineering risks, health risks, information risks and so forth. According to Frank Knight's seminal remarks in *Risk, Uncertainty and Profit* (1921, reissue 2005), we should preliminarily grasp that "risk, as loosely used in everyday speech and in economic discussions, really covers two things which, functionally at least, in their causal relations to the phenomena of economic organization, are categorically different." Those two things are proper risk as "a quantity susceptible of measurement" or "measurable uncertainty," and risk that may be difficult or impossible to quantify, referred to as proper uncertainty. For example, when dealing with the safety factors of structural engineering, *e.g.*, the safety structures of buildings, scholars distinguish between sources of failure amenable to probabilistic assessment, such as poor qualities of materials and higher loads than those foreseen in the project, and uncertain factors such as human error, potentially unknown failure mechanisms, or the imperfect theory of the failure mechanism in question, *i.e.*, proper uncertainty.

Although these three approaches to the notion of risk are intertwined, let us restrict our focus to cases of strict risk and the ways scholars address the challenges of proper uncertainty. A fruitful illustration is offered by the nuclear industry and how, in the 1950s and 1960s, engineers designing nuclear reactors intended to keep the probability of accidents as low as possible, although they did not have any methodology to determine such probabilities. In fact, modern probabilistic risk assessment developed only in the late 1960s and early 1970s, culminating with the 1975 Rasmussen report. In the phrasing of Neelke Doorn and Sven Hansson (2011: 155), "the basic methodology used in this report is still used, with various improvements, both in the nuclear industry and in an increasing number of other industries as a means to calculate and efficiently reduce the probability of accidents." In a nutshell, this probabilistic approach aims to single out the undesirable events to be covered by the analysis, so as to pinpoint the accident sequences that may lead to the occurrence of adverse events as well as the probability of each event in the sequence.

In light of early versions of probabilistic risk assessment, two "improvements" in today's approach should be mentioned. First, experts do not aim at establishing the overall probability of a serious accident but rather, the weaknesses in the safety system, by ranking the accident sequences in connection with the probability of their occurrence. Then, probabilities are not conceived of as "unbiased predictors of occurrence frequencies that can be observed in practice," but "as the best possible expression of the degree of belief in the occurrence of a certain event."[1] This is why, back to the view of

[1] See the definition of the "probabilistic model code" proposed by the Joint Committee on Structural Safety (JCSS 2001: 60).

Doorn and Hansson, experts of probabilistic risk assessment "in the nuclear industry have largely given up the original idea that the outputs of probabilistic analysis of event sequences in nuclear reactors could be interpreted as reasonably accurate probabilities of various types of accidents. Instead, these calculations are used primarily to compare different event sequences and to identify critical elements in these sequences" (Doorn and Hansson 2011: 157).

Such constraints emphasize the critical limits of risk analysis, especially when we are confronted with new and untested technologies and, thus, a lack of data. The empirical basis of probabilistic models necessarily hinges on events that are common enough to let scholars collect data about their occurrence and, yet, probabilities of unusual events may be the most relevant ones in risk analysis. Although further methods have been developed to assign probabilities to rare events, such as extreme value analysis, distribution arbitrariness or boot-strapping methodologies, such approaches may fall short in coping with the unpredictable behaviour of autonomous machines. For example, "boot-strapping techniques still require sufficiently long data records and a careful analysis of the influence of data sampling uncertainties" (Doorn and Hansson 2011: 158). Moreover, certain scholars reckon that measurable risks can hardly be assigned to human reactions vis-à-vis novel or experimental technologies. Rather than hinging on probabilities, the focus should be on qualitative or human-centred approaches, so as to delimit the sphere of uncontrollable uncertainties by singling out new types of human failure (Mosneron-Dupin et al. 1997).

Leaving aside further risk analysis approaches, such as the "partial safety factors" proposed by Isaac Elishakoff (2004), we may wonder how the advancement of robotics technology affects the field. As mentioned in the introduction to this chapter, contractual obligations and rights concerning the design, construction and use of robots are strictly related to the level of risk and predictability of their behaviour. Whereas in *The Laws of Man over Vehicles Unmanned* (2008), Brendan Gogarty and Meredith Hagger claim that "determining fault in complex software and hardware is already difficult" (*op. cit.*, 123), let us consider three different scenarios.

First, we have the da Vinci surgical system that, according to the website of its manufacturer, Intuitive Surgical, "enables surgeons to perform delicate and complex operations" such as prostatectomy procedures, "through a few tiny incisions with increased vision, precision, dexterity and control." Work in the *Mechanical Failure Rate of da Vinci Robot System* shows that only 9 out of 350 procedures (2.6 %) could not be completed due to device malfunctions (Borden et al. 2007). Likewise, in *Device Failures Associated with Patient Injuries During Robot-Assisted Laparoscopic Surgeries* (2008), Andonian et al. affirm that only 4.8 % of the malfunctions that occurred in a

New York urology institute from 2000 to 2007 were related to patient injury. What happens, from a legal viewpoint, in such cases where these artificial doctors do not properly work is examined next in Sect. 4.2.

The second scenario is illustrated by the mishap rate of the unmanned aerial vehicles such as the US Air Force's RQ-1 Predator or the US Army's RQ-2 Pioneer. According to the US Air Force's catalogue, we should distinguish three classes of accidents:

(a) Class A mishaps that include the destruction of $ 1 million in property, loss of a Department of Defence aircraft, or a human casualty resulting in loss of life or permanent disability;
(b) Class B mishaps that involve a $ 200,000–$1 million in property damage, human casualty leading to partial disability or three or more hospitalized personnel; and
(c) Class C mishaps that finally concern a $ 20,000–$ 200,000 in property damage or non-fatal injury leading to a loss of time at work.

By 2005, the level of risk for UAVs was much higher than for traditional aircrafts. When compared to manned aviation, the US Air Force's RQ-1 Predator had 32 times as many accidents per flight-hour, the US Navy's RQ-2 Pioneer more than 300 times and the US Army's RQ-5 Hunter approximately 60 times as many as traditional manned aviation. Accordingly, Peter Singer estimates in *Wired for War* (2009) that notwithstanding technological advancement, training or safer operations under peacetime conditions, UAV security "needs to improve by one to two orders of magnitude to reach the equivalent level of safety of manned aircraft."

Such poor figures certainly characterize the civilian use of UAVs as well. Remarkably, the American National Transportation Safety Board ("NTSB") examined three cases of domestic UAV mishaps between 2006 and 2008. In the wording of Geoffrey Rapp's work on *Unmanned Aerial Exposure* (2009), let us see what occurred in one of these cases:

> In April 2006, a Predator UAV used by the United States Customs and Border Protection Service crashed into the Arizona desert when its operators turned off its engine. When one of the Predator's two ground control stations locked up during flight, its operator switched to the other station but neglected to 'align consoles,' inadvertently cutting off the platform's fuel supply. As the UAV lost power during flight, it began to 'shed electrical equipment to conserve electrical power' [according to the NTSB report].
>
> Although no one on the ground was injured, 'the accident didn't help the industry's reputation' (Stew Magnuson). The UAV glided as close to 100 feet from two homes before striking the ground; homeowners heard the crash and thought a bomb had exploded. The NTSB attributed the crash to inadequate surveillance of the program, pilot error, and inadequate maintenance procedures performed by the manufacturer.

> … Accidents like this have thankfully caused no injuries to date, but widespread use of UAVs in the domestic setting would inevitably produce casualties and property loss as a result of crashes or objects falling from airborne UAVs (G. Rapp, *op. cit.*, 628–629).

The final scenario has to with the point of view of insurance companies and risk management. Such companies are third parties to contracts that either pay out when someone else commits a tort against the insured, or cover losses sustained by the insured against a premium, *i.e.*, the factor through which the sum to be charged for a certain amount of insurance coverage is established. Consider the civilian employment of UAVs and how different uses of such technology are covered by policies such as business or pleasure, commercial or industrial aid. According to Geoffrey Rapp (2009), one commercial UAV imagery company, Moire Inc., "carries $2 million in liability insurance and invites customers to request categorization as 'Additional Insured' under its policy" (*op. cit.*, 647). Moreover, when UAVs are employed for scientific purposes, the premium "has been nearly 85 % of the cost of operation per flight hour" and with respect to hull insurance policies, their cost "has been estimated to reach 2 % of UAV replacement value, plus 0.5 % of ground station replacement value and $30,000 per UAV mission" (*ibid.*).

What these examples of insurance costs suggest is the need to grasp the panoply of robotics applications and their impact on clauses and conditions of contractual obligations in light of a spectrum. At one end, we find a number of reasonable safe and controllable robots that, due to the well-established quantifications of the probability of events, their consequences and costs, do not raise particular challenges to traditional risk assessment analysis or the risk management of insurance companies. At the other end of the spectrum, the progressively unpredictable behaviour of robots raises problems of proper uncertainty, rather than quantifiable risk in the construction and use of these machines. The more we widen the settings and goals of robotic programs, the more we will be dealing with growing amounts of complexity, so that the risks emerging will exponentially increase as a consequence of robotic behaviour. Although we do not have to accept Curtis Karnow's idea that the advancement of robotics will end up in a failure of legal causation as discussed above in Sect. 3.5, it is likely that in the field of contracts, the growing autonomy of robots will affect basic concepts such as foreseeable harm, individual negligence or fault. By considering cases of reasonable safe and controllable robots as seen in the next section, we set the background for the analysis of a new generation of robots that fall within the loopholes of today's legal framework and are further discussed in Sect. 4.3.

4.2 The Artificial Doctor

This section focuses on the case of the da Vinci surgical system as an example of how a significant number of robotic applications do not challenge today's legal framework on matters of liability for the behaviour of such machines. This does not mean, of course, that robotic surgery does not raise certain critical issues. For example, in *Predicting the Long-Term Effects of Human-Robot Interaction* (2011), Edoardo Datteri points out "cases of harmful (occasionally fatal) events brought about by negligent use of medical robots behaving normally." Although da Vinci surgical systems may reduce hospital stays by about one-half and hospital costs by about a one-third, there is the risk of "negligence due to poor training with the robotic system: surgeons [are] not given enough time and resources to learn to use the robot properly, … whereas surgeons with extensive robotic experience declare that it takes a minimum of 200 surgeries to become proficient at the Da Vinci" (*op. cit.*) In *Robotic Surgery Claims on United States Hospital Websites* (2011), Linda Jin et al. argue that the use of such robots appears more as a marketing tool to attract patients than a medical system to improve their care. Through a systematic analysis of 400 randomly selected US hospital websites in June 2010, Jin et al. reckon that "forty-one percent of hospital websites described robotic surgery. Among these, 37 % presented robotic surgery on their homepage, 73 % used manufacturer-provided stock images or text and 33 % linked to a manufacturer website. Statements of clinical superiority were made on 86 % of websites, with 32 % describing improved cancer control and 2 % described a reference group. *No hospital website mentioned risks. Materials provided by hospitals regarding the surgical robot overestimate benefits, largely ignore risks and are strongly influenced by the manufacturer*" (*op. cit.*, italics added). Significantly, the Los Angeles Times published an article on 17 October 2011 by Amber Dance, summing up some of these concerns: "Robotic surgery grows, but so do questions. The Da Vinci system is now in 2,000 hospitals. But there's concern that hands-on surgery still has advantages."

From a legal viewpoint, however, both the design and construction of such robots, as well as their employment in 2,000 hospitals, do not seem particularly challenging. As shown by the case, *Mracek v. Bryn Mawr Hospital*, discussed below in Sect. 4.2.2, the current legal framework concerning liability issues for harm caused by the malfunctioning of electronic devices can properly address harms induced by robotic breakdowns. Yet, such cases of liability do not only have to do with clauses and conditions of contracts established between private persons; namely, in the case of the da

Vinci surgery system, the designer and producer of such robots, Intuitive Surgical, and the user of these machines, such as hospitals and natural (rather than artificial) doctors. In fact, the use of such robots may concern the rights of third parties as well as obligations imposed by the state so as to compensate for any damages done by wrongdoing. Therefore, how clauses and conditions of contracts may involve rights and interests of third parties and, *vice versa*, how the legal protection of third parties may affect contractual rights and obligations are examined in the next Sect. 4.2.1. The focus then is on the claims of a third party, Roland C. Mracek, filing suit against both the producer of the da Vinci surgery system and one of its users, the Bryn Mawr hospital in Philadelphia, due to the malfunctioning of a da Vinci system, as explored in Sect. 4.2.2.

4.2.1 Parties, Counterparties and Third Parties

The employment of robotic applications concerns clauses and conditions established by the parties to a contract as well as the rights and interests of third parties. In addition to insurance companies as third parties covering either losses sustained by the insured or paying off when the insured harms another party, consider what occurred to certain homeowners in the Arizona desert in April 2006. These homeowners heard a Predator UAV gliding as close as 100 ft to their houses before striking the ground and making them think that a bomb had exploded. Luckily no injuries were caused by the UAV.

Two types of obligations must be distinguished concerning designers, producers and users of robots that may damage third parties. Some obligations depend on a voluntary agreement between private persons, others are generally imposed against the will of the agent. This type of extra-contractual responsibility includes cases of intentional wrongdoing, negligence-based liability and strict liability. What common law lawyers sum up with the term of tort, may raise forms of apportioned responsibility between the parties to a contract as discussed above in Sect. 2.2.

Let us now view how this complex set of notions works in practice by taking into account a prostatectomy operation by the da Vinci robot. For example, in *Mracek v. Bryn Mawr Hospital*, we have to distinguish four levels of analysis:

(a) The parties to the contract, that is, Intuitive Surgical and the Bryn Mawr Hospital, that determine the conditions for the use (and maintenance) of a da Vinci surgery system;

(b) The insurance company as a third party to that contract on a voluntary basis (although we will examine cases of compulsory insurance in the next section);
(c) Another third party who voluntarily underwent surgery with the da Vinci system, namely, the patient Roland Mracek and his contract with the Bryn Mawr hospital; and
(d) A tort liability suit filed by the patient as Mracek claims to have suffered unwarranted damages caused by both the parties to the contract (*sub a*), that is, Intuitive Surgical and the Bryn Mawr Hospital.

Contractual parties, when establishing the clauses and conditions of their agreement (*sub a*), will thus have to pay attention to the obligations imposed by the state in order to compensate for unjust damages (*sub d*). Consider contracts of software developers that often establish clauses of strong liability limitations and even exemptions for damages caused by their products. *Vice versa*, reflect on the case of US federal contractors that pursuant to 28 U.S.C. § 2671, know that clauses of immunity that protect their contractual counterparties do not extend to them as seen above in Sect. 3.5. In *Mracek v. Bryn Mawr Hospital*, it is noteworthy that one of the defendants, the Bryn Mawr hospital, was dismissed from the suit by court order. Only Intuitive Surgical, the designer and producer of the robot, had to defend itself by showing that the da Vinci robot did not cause any unjust damage. In order to understand how claims of third parties (*sub d*) may affect conditions and clauses of contracts (*sub a*), we shall focus on the different ways the apportioned liability between the parties to a contract depends on three types of extra-contractual responsibility.

First, liability can be ascribed to the tortfeasor for wrongful conduct because that person intended to do harm. Contemplate the case of a doctor who voluntarily causes harm to a patient through the use of the da Vinci robot system. Whereas, in criminal law, the hypothetical of an intentional tort brings us back to the second step of the phenomenology of *Picciotto Roboto*, see above in Sect. 3.4.2, in the civil (as opposed to the criminal) law field, such a wrongful intention severs the link between claims of extra-contractual liability (*sub d*) and previous contractual obligations (*sub a*). It is clear that the producer of the robot is not to be held liable for the conduct of the user of the machine.

Second, there is the opposite case of strict liability, or liability without fault, invoked when the conduct of the tortfeasor is not blameworthy. Regardless of the absence of any illicit or culpable behaviour, individuals are held liable for damages caused by their own dangerous activities or the behaviour of other agents in the legal system. In the case of strict product liability, it follows that claims of extra-contractual responsibility (*sub d*) can overrule contractual agreements for the design, construction and supply of such a product (*sub a*). At times, the producer, rather than the user, of the robot will have to show that there is no evidence that the machine did not properly work.

Finally, liability can be based on negligence or lack of due care, *e.g.*, when a reasonable person fails to guard against foreseeable harm. As mentioned above in Sect. 3.5, strict liability rules do not prevent additional individual liability for careless conduct. Furthermore, a negligence claim may stand even in the absence of a defect under strict liability norms. The link between extra-contractual liability (*sub d*) and contractual obligations (*sub a*) hinges, therefore, on the circumstances of the case, so as to determine whether the user or the producer was negligent.

In light of this general framework, let us deepen how robotic applications affect clauses of civil (as opposed to criminal) responsibility. In this context, we can set aside cases of intentional torts as well as crimes of intent: as shown by the second step of the phenomenology of *Picciotto Roboto* in Sect. 3.4.2, these hypotheticals end up in the class of plain cases. The focus rather should be on strict liability rules and cases of negligence in the civil law field and how the burden of proof is allocated in such cases. Regardless of the differences between common and civil law systems, discussed more thoroughly below in Sects. 5.2 and 3, this complex set of notions and procedures can be illustrated with *Mracek v. Bryn Mawr Hospital*. In this case, the patient/plaintiff alleged that the da Vinci robot caused damage arising out of strict product and malfunction liability, negligence and breach of warranty. The reasons why plaintiff finally lost his case introduce a new class of plain cases in the laws of robots. The general agreement depends on the fact that there are a number of reasonably safe and controllable robots out there.

4.2.2 Producers, Users and Patients

Something went wrong with the surgical removal of a part of Roland Mracek's prostate at the Bryn Mawr Hospital in Philadelphia on 9 June 2005. According to the plaintiff, liability for erectile dysfunction and groin pain following from the medical procedure should be imposed on both the producer (Intuitive Surgical) and the user (Bryn Mawr Hospital) of the da Vinci surgery system. Such a machine would have caused damages, first of all, due to its own malfunctioning, so that the producer of the robot should be held strictly liable. In the phrasing of § 402A of the Restatement (Second) of Torts in the US, strict liability is imposed "not only for injuries caused by the defective manufacture of products, but also for injuries caused by defects in their design." In such cases, the burden of proof falls on the plaintiff who has to prove that the product was defective; that such defect existed while the product was under the manufacturer's control; and, moreover, the defect was the proximate cause of the injuries suffered by the plaintiff. Both the standards and burdens of proof required by § 402A of the Restatement (Second) of Torts apply to liability claims for breach of warranty as well.

Plaintiff's second claim has to do with provisions of strict malfunction liability. Responsibility can be imposed although the plaintiff is not able to produce direct evidence on the defective condition of the product or the precise nature of the product's defect. Rather, the plaintiff is to demonstrate that defect through circumstantial evidence of the occurrence of a malfunction, or through evidence eliminating both abnormal use of the product and reasonably secondary causes for the accident.

Finally, responsibility for civil (as opposed to criminal) negligence hinges on the duty to conform to a certain standard of conduct. Here, the plaintiff has to prove that defendants breached that duty, thereby provoking an injury and an actual loss or damage to the plaintiff.

Interestingly, Mracek did not submit any expert report to support or corroborate his claims. In the wording of the District Court, the plaintiff's argument was that the asserted defect of the robot was "obvious enough to be ascertainable by the average juror without speculation." More particularly,

> Mracek contends that an expert report is not necessary because the surgeon who performed his operation, Dr. McGinnis, will testify at trial concerning not only his pre- and postoperative medical condition, but also the malfunction of the da Vinci robot. Mracek maintains that the defect of the surgical robot is obvious because all of its component parts shut down after repeatedly flashing "error" messages, and then was not able to be restarted once the surgery commenced. Mracek argues that it is not necessary for him to produce an expert report for a finding of an obvious defect, as such a defect is not beyond the purview of a layperson when presented with this factual record (District Court of Philadelphia, Judge R. Kelly, *case 08-296* from March 11, 2009, *cit.*, 6).

Although "absence of expert testimony is not fatal to a products liability case," this principle does not typically apply to such complex machines as the da Vinci robot. All in all, this is why Mracek lost the case. According to the court, the plaintiff failed to support his case without an expert report, because he could not establish either a defect of the robot or a causal link between the problems with the robot and the plaintiff's damages under strict liability rules. Likewise, under the malfunction theory of strict products liability, the plaintiff did not offer any evidence so as to eliminate reasonable secondary causes, nor did he produce any genuine issue of material fact regarding elements of negligence that could be given to a jury. Therefore, the court granted the defendant's motion for summary judgment against Mracek in 2009. Under US Federal Rule of Civil

Procedure 56(c), summary judgment is to be granted "if there is no genuine issue as to any material fact and the moving party is entitled to judgement as a matter of law."

The Court of Appeals confirmed the District Court's judgment in 2010, with Justices Scirica, Barry and Smith rejecting Mracek's argument that the District Court improperly granted summary judgment on his strict malfunction liability claim.[2] The court reasoned that the trial court's decision "was proper because he [Mracek] failed to demonstrate a genuine dispute of material fact. Most importantly, there is no record evidence that would permit a jury to infer Mracek's erectile dysfunction and groin pain were caused by the robot's alleged malfunction" (*op. cit.*, 5). As the plaintiff cannot depend upon simple conjecture or guesswork and has to introduce "evidence from which a rational finder of fact could find in his favour," the Court of Appeals confirmed the trial court's grant of summary judgment. Four months later, Mracek filed a petition for a writ of certiorari before the Supreme Court, which was distributed for conference in September, and a few days later, on 4 October 2010, denied.

After the set of plain cases on crimes of intent as examined in Sect. 3.4.2, the *Mracek vs. Bryn Mawr Hospital* case illustrates a further class of "general agreement in judgments as to the applicability of the classifying terms" (Hart 1994: 123). On one hand, Mracek's case seems plain because of the lack of evidence. On the other hand, as a genuine dispute of material fact, in the phrasing of the Court of Appeals, we can imagine an alternative outcome of the case, *i.e.*, the plaintiff could prove the causal link between the behaviour of the robot and his erectile dysfunction. Yet, traditional notions of the law, such as proximate or reasonable secondary causes, negligence or breach of warranty, would still be at work. The reason why the behaviour of the da Vinci system does not affect how lawyers grasp individual liability in these cases, hinges on the controlled settings of the operational theatres that delimit the conduct of the machine: its mechanisms and properties do not look more intricate than the complex analysis of scientific experts in other fields of the law as raised above in Sect. 3.5. After crimes and torts, both of which depend on the "wrongful" conduct of humans, this class of plain cases referring to hypotheticals of strict malfunction rather than strict product liability,

[2]Mracek's appeal did not concern his previous claims on strict product liability, negligence and breach of warranty. In *Unmanned Vehicles and US Product Liability Law* (2012), Stephen S. Wu addresses further cases where "defendants were entitled to summary judgement because the plaintiffs failed to introduce evidence in opposition to summary judgement showing that the system was defective." Among such cases, see *Jones v. W + M Automation*, 818 N.Y.S. 2d 396 (App. Div. 2006), appeal denied, 862 N.E. 2d 790 (N.Y. 2007); and *Payne v. AAB Flexible Automation*, 96–2248, 1997 WL 311586 (8th Cir. Jun. 9, 1997).

represents the first end of the spectrum of robotic applications, namely, machines that are reasonably safe and controllable.

However, it is not so difficult to conceive of more complex cases. Let us dwell on the Bryn Mawr Hospital and imagine the more than realistic scenario of an artificial agent working at that hospital, scheduling the appointments of patients. The agent checks priorities for surgeries performed by the da Vinci surgery system and alerts maintenance staff and so forth. This robot suggests we are dealing with a proper agent, rather than a simple tool of human interaction. There are already, after all, a number of such agents that terminate or renew Medicaid programs, food stamps and other welfare schemes, by enrolling "applicants directly into benefits programs without review or critique by human operators" (Chopra and White 2011: 195). Furthermore, by widening the set of parameters and conditions regulating the behaviour of the robot, *e.g.*, machines operating in open environments, it is likely that the level of risk and proper uncertainty arising from the use of such machines will severely impact basic tenets of the law and, more particularly, the field of contracts. In *Agent Technology: Computing as Interaction* (2005), Michael Luck et al. draw attention to a number of possible candidates for a new generation of legal hard cases, such as "simulation and training applications in defence domains; network managements in utilities networks; user interface and local interaction management in telecommunication networks; schedule planning and optimisation in logistics and supply-chain management; control system management in industrial plants," up to simulation modelling "to guide decision makers in public policy domains" (*op. cit.*, 50).

Here, the legal challenges of robotics in the field of contracts can be illustrated with a class of machines that may negotiate deals, accept bids, send offers and establish rights and duties of their own. Contrary to the controlled settings of the da Vinci system, the class of trading artificial agents may affect basic notions and ways of legal reasoning in three different ways. First, such machines can successfully be used to carry out complex business transactions and, yet, their behaviour, at times, suggests troubling parallels with the greediness of human speculators. Second, these robots are traditionally presented as instruments of human interaction and, still, an increasing number of scholars reckon that such robots should be conceived as new actors in today's legal systems. Finally, strict liability rules currently apply to robots and, nevertheless, such artificial agents suggest new forms of accountability and responsibility for the behaviour of others in both contracts and tort law. Therefore, let us proceed with the analysis at the opposite end of the spectrum represented by the reasonable safe and controllable robot examined in this section. Matters of risk and, moreover, of proper uncertainty as at the other end of the spectrum, are at stake with a new generation of robo-traders.

4.3 Robo-Traders

Work in artificial trading agents has been cutting edge in the past few years. Along with contributions to the trading agent competition ("TAC")-context, such as Seong Jae Lee et al. in *RoxyBot-06: An (SAA)2 TAC Travel Agent* (2007), we can mention the works of Jeffrey Mackie-Mason and Michael Wellman in *Automated Markets and Trading Agents* (2006), Michael Wellman, Amy Greenwald and Peter Stone in *Autonomous Bidding Agents* (2007), Giovanni Sartor in *Cognitive Automata and the Law* (2009), Samir Chopra and Laurence White in *A Theory for Autonomous Artificial Agents* (2011). Whereas, most of the time, these works focus on software agents, rather than robots interacting in the real world, such machines raise some common issues. On one hand, their behaviour and decisions can be unpredictable and risky, as shown by robotic experiments in double auction markets throughout the past decades. Here, the traditional legal viewpoint considers robots simply as tools or means of human interaction, which means that strict liability rules apply to humans as principals of the machine. On the other hand, there is a number of reasons why some of these robots should be deemed as proper agents rather than tools of human interaction: such machines can be extremely efficient in establishing rights and obligations between humans that delegate to them complex cognitive tasks. As a result, today's strict liability rules raise the threat that people think twice before employing robots that may provide "services useful to the well-being of humans" (UN World Robotics 2005). Richard Posner summarizes this popular stance when claiming that the best method of accident control is to scale back the activity (Posner 1973: 180).

This section sheds light on the legal challenges of robotics through a case study in the field of artificial trading agents. Next, attention is paid to the robotic experiments in double auction markets in Sect. 4.3.1 in order to illustrate the pros and cons of such technological applications. The first laboratory double auction in markets, where buyers and sellers submit bids and offers in any order, was reported by Vernon Smith's classic paper *An Experimental Study of Competitive Market Behaviour* (1962). Some thirty years later, robot tournaments were conducted at the Santa Fe Institute and, in the early 2000s, an Automated Trading project in robots, trading in auction markets, was sponsored by the University of Pennsylvania and Lehman Brothers. This case study is deepened in Sect. 4.3.2: the focus is on the traditional legal viewpoint that holds individuals responsible for the use of such robot traders according to the rules that apply to users as principals of these machines. By showing how today's strict liability rules fall short in coping with certain legal challenges of robot traders in Sect. 4.3.3, the aim of Sect. 4.4 is to provide a more fruitful guide to a new generation of hard cases in the legal domain.

4.3.1 Artificial Greediness

The baseline for all robot archetypes in double auction markets is given by the Zero Intelligence ("ZI") agents. These robots are rudimentary in that they are oblivious to their environment and do not control the timing of their actions: ZI agents even lack the capability of taking action so as to compensate for their inability to respond to the environment. As Ross Miller argues in his telling article *Don't Let Your Robots Grow Up to Be Traders* (2008), a ZI agent is a robot programmed to simply "generate bids and offers selected randomly from a uniform distribution subject only to the constraint it cannot 'deliberately' lose money." However, if ZI agents are certainly rudimentary, they also achieve sophisticated goals as outperforming untrained human traders in double auction experiments. Moreover, the performance of ZI agents in shopping around or planning ahead can be improved, so that according to Miller, "the design of a special-purpose agent that can trade in the simple asset markets… as well as, if not better than, humans seems clearly within grasp" (*op. cit.*).

Interestingly, since the robot tournaments at the Santa Fe Institute in 1990, scholars have programmed ZI agents in order to replicate human double-oral auctions, showing that markets populated only by such robots have the tendency of human markets to generate average prices and quantities of what economists traditionally present as a "competitive equilibrium." As Shyam Sunder affirms in *Markets as Artefacts* (2004), computer simulations have demonstrated "that allocative efficiency – a key characteristic of market outcomes – is largely independent of variations in individual behaviour under classical conditions." This ability of ZI agents to achieve a high level of allocative efficiency when determining average prices and quantities of goods exchanged in a market can be grasped with Friedrich Hayek's idea that in certain fields of social interaction, such as pacts and contractual obligations, "intelligence" emerges from the rules of the game rather than individual choices. Yet, a lot of problems arise when addressing the subtleties of markets containing intelligent agents such as humans. Work on robot trading in auction markets as the Automated Trading project, sponsored by the University of Pennsylvania and Lehman Brothers, showed relevant failures as to programming robot traders capable of effectively speculating against (smart) humans. It is noteworthy that this project was finally suspended in 2005, that is, 3 years before Lehman Brothers' own collapse…

In addition, the complexity of tackling multiple actions occurring synchronically in time far exceeds the capabilities of ZI agents. This circumstance reduces the allocative efficiency of the market and leads to a rudimentary bubble and crash scenario, where traders act without regard of the effects of future supply. As in real life bubbles, agents are overwhelmed by the complexity of the environment, thereby appearing extremely inexperienced. This analogy has suggested that experiments with the random-bidding strategy employed by such robots can clarify how real life bubbles form. As stressed by Miller (2008), "the bubble in Internet and other technology stocks that formed at the end of the 1990s may have been partially rooted in market participants' inability to properly anticipate the future supply of stock in Internet companies." Similarly, others argue "that some of the financial troubles of late 2009 may have been caused by the involvement of such agents operating without human supervision and at speeds not amenable to human understanding or intervention" (Chopra and White 2011: 7).

The parallel between the greediness of human speculators and the eagerness of ZI robots to trade, however, does not mean that such artificial agents should not be preferred to humans in certain market operations, *e.g.*, when speed is valued over intelligence. Moreover, there are a number of robotics applications and, generally speaking, of autonomous artificial agents that do not raise such a level of risk when, say, individuals bid, buy or book. Suffice it to mention today's routine interaction with eBay bidding agents, iTunes store agents, Amazon's website bots, or the common airline booking system that through "yield management techniques," determine prices according to how crowded the flight is and so forth. By opening up new ways of "making business as usual," *e.g.*, granting authority to the artificial agent so as to let it act on an individual's behalf when dealing with third parties, we should pay attention to how the law aims to govern such business. For example, we may agree with the American Law Institute and Commissioners of the Uniform State Laws that contracts made by electronic agents should be considered valid, although no action or knowledge of any human being may be involved. Still, this approach leaves open the question of whether humans are bound by every decision of a robot and which human party would be bound by such decision: the designer/implementer of the robot, its user, the operator or the principal?

4.3.2 *The Robot and the Principal*

Rights and obligations established by robots can be interpreted through the traditional legal viewpoint as examined already with the artificial doctor. Strict liability rules should in fact govern the behaviour of robots, binding those humans on whose behalf they act, regardless of whether such conduct was planned or envisaged. In the US, for example, the E-SIGN statute and the 1999 attempt to amend the Uniform Commercial Code with a Uniform

Computer Information Transactions Act ("UCITA") illustrate this approach. On the one hand, 15 U.S.C. § 7001(h) provides that a contract "may not be denied legal effect, validity or enforceability solely because its formation, creation or delivery involved the action of one or more electronic agent so long as the action of any such electronic agent is legally attributable to the person to be bound." On this basis, in *Spiders and Crawlers and Bots* (2002), Jeffrey Rosenberg claims that "a robot that enters into a clickwrap agreement, either by clicking on an 'I accept' button, or disregarding the express protocol set forth in a robot exclusion header, binds the person who designed and implemented the robot."

On the other hand, section 107 (d) of UCITA establishes that "a person that uses an electronic agent that it has selected for making an authentication, performance or agreement, including manifestation of assent, is bound by the operations of the electronic agent, even if no individual was aware or reviewed the agent's operations or the results of the operation." Likewise, the Unicitral document enclosed in the proposal of the UN Convention Electronic Communications in International Contracts Documents states that "general principles of agency law (for example, principles involving limitation of liability as a result of the faulty behaviour of the agent) could not be used in connection with the operation of such systems. The Working Group reiterated its earlier understanding that, as a general principle, the person (whether a natural person or a legal entity) on whose behalf a computer was programmed should ultimately be responsible for any message generated by the machine… As a general rule, the employer of a tool is responsible for the results obtained by the use of that tool since the tool has no independent volition of its own."

Summing up the outcomes of the robots-as-tools approach, we consequently have:

(a) Robot *R* acting on behalf of the principal *P*, so as to negotiate and make a contract with the counterparty *C*;
(b) Rights and obligations established by *R* directly bind *P*, since all the acts of *R* are considered as acts of *P*;
(c) *P* cannot evade liability by claiming either she did not intend to conclude such a contract or *R* made a decisive mistake;
(d) In case of the erratic behaviour of *R*, *P* may claim damages against the designer and producer of *R*. However, according to the mechanism of the burden of proof, *P* will have to demonstrate that *R* was defective and that such defect existed while *R* was under the manufacturer's control; and, moreover, the defect was the proximate cause of the injuries suffered by *P*.

Although the traditional outlook may fit under certain circumstances, the robots-as-tools approach is flawed for three reasons. First, it is likely that most of the time, humans will delegate to autonomous and even smart robots complex cognitive tasks, such as acquiring knowledge for decision-making. Consequently, it is difficult to accept the traditional idea that robots are mere tools of human interaction and, moreover, that rights and obligations established by robots would be directly conferred upon humans (*sub b*), because the principal wanted the specific content, or agreement, of the contract made by the artificial agent. Rather, rights and obligations are conferred onto humans because they delegate to the robot the authority to act on their behalf.

Second, from the fact that *P* delegates to *R* (*sub a*), it does not follow that the legal effects of the behaviour of *R* should necessarily fall upon *P* (*sub b*). Admittedly, the robot's counterparty *C* should be allowed to expect, in good faith, that the machine really means what it declares, *e.g.*, a contractual offer, when negotiating with robot *R*, so that *P* cannot evade liability by claiming she did not intend to conclude such a contract (*sub a*). However, humans should not be able to avoid the usual consequence of robots making a decisive mistake, *i.e.*, the annulment of a contract, when *C* had to have been aware of a mistake that due to the erratic behaviour of the robot, clearly concerned key elements of the agreement, such as the market price of the item or the substance of the subject-matter of that contract. Here, it seems reasonable to expect that the humans involved in such transactions should be bound by the interpretation of the behaviour of the robot that usually applies to the circumstances of the case according to existing conventions of business and civil law.

Third, the robots-as-tools approach appears unsatisfactory when responsibility (and risk) must be distributed between, say, operators and users as principals of the robot. Whereas the traditional approach ends up in a Hegelian night where all kinds of responsibility look grey, operators and users of robots should be held accountable in accordance with the different errors of the machine and the circumstances of the case. In fact, the erratic behaviour of the robot can concern not only software and hardware malfunctioning, or errors of specification as mentioned above, *e.g.*, errors concerning the substance matter of a contract. In the phrasing of Chopra and White (2011: 46), we should take into account "induction errors, where a discretionary agent incorrectly induces from contracts where the principal has no objections to a contract the principal does object to." Aside from a further hypothetical of liability involving the manufacturers of the artificial agent, we should also distinguish cases where operators and users of the robot coincide and cases where operators allow users to use the machine, so as to deal with third parties. Nine possible cases follow as a result: the legal variables of this section are illustrated with Table 4.1. "Yes" and "no" refer

Table 4.1 What the approach to robots-as-tools lacks

Erratic robot	Specification	Induction	Malfunction
Human operator	Yes	Yes	Sometimes no
Human user	Yes	No	Sometimes no
Third parties	No	No	Sometimes yes

to whether or not human operators, users or third parties should be held accountable for the erratic conduct of the machine:

In *A Legal Theory for Autonomous Artificial Agents* (2011), Chopra and White examine this complex scenario by further considering the theories of the unilateral offer, of the objective intention, and so forth (*op. cit.*, 45–50). Here, it suffices to pay attention to the three rows of Table 4.1. The first set of cases concern legal responsibility of the human operator for the erratic behaviour of the robot due to specification errors, induction mistakes or the malfunction of the machine. Compared with the strict liability approach, according to which operators might be liable under all circumstances, it is arguable that such an operator should not be accountable for malfunctions of the machine that are obvious to users and third parties. In the wording of *A Legal Theory for AAAs*:

> An example of the first kind of transaction occurs when the principal is the operator of a shopping website (such as Amazon.com), the agent is the website interface and backend, and the third party is a user shopping on the website. The contract is formed between the principal and the third party…
>
> When the principal is the agent's operator, specification and induction errors will be less obvious to third parties than to principal/operators, and therefore the principal/operator will normally be the least-cost avoider of the loss. Where, for example, because of specification or induction error, a book is advertised very cheaply, the third party may simply understand the price to be a "loss leader" rather than the result of an error… In the case of malfunction it may be obvious to the third party, because of other indications, that a particular price is the result of error… Therefore, often, the least-cost avoider of malfunction errors will be the third party.
>
> With the agent understood as a mere tool, the principal would be liable for all three types of error in all cases. This approach would not be efficient where the third party is the least-cost avoider of the risk, as in many cases of malfunction error (Chopra and White, *op. cit.*, 46–47).

Vice versa, we can imagine cases where the principal is the user, rather than the operator, of the artificial agent. After all, this is what occurs on eBay, where individuals use the auction website's proxy bidding system so as to enter a contract with a third party:

> In this case, as in the operator as principal case, the risk of specification errors should normally fall on the principal, that is, the user of the agent. However, the risk of induction errors should normally fall on the operator of the agent (who has control over the agent's design and operation). The risk of malfunction errors will

> often most fairly fall on the third party, for the reasons given in discussing the operator as principal case.
>
> Under the "agent as mere tool" solution, the user/principal would be primarily liable for all three types of error, incorrectly allocating the risk of induction and malfunctions error in particular (Chopra and White, *op. cit.*, 48–49).

The final row of Table 4.1 concerns responsibility of third parties for the threefold erratic behaviour of robots. As stated in this section, the traditional legal stance falls short in coping with the accountability of those who have to be aware of, say, a mistake of the robot due to its erratic behaviour. Aside from the allocative efficiency of such no-fault responsibility rules, there is the risk that strict liability policies can dissuade humans from employing robots at all. Is there a feasible way out of the *cul-de-sac* that characterizes the robots-as-tools approach?

4.3.3 A New Agent in Town

It makes a lot of sense to conceive (certain types of) robots as proper agents in the field of contracts, that is, granting them the authority to act on an individual's behalf when dealing with third parties. Such a perspective prevents certain key flaws of the robots-as-tools approach, since the legal agency of the robots makes it clear that humans do delegate crucial cognitive tasks to these machines. We can establish individual responsibility for the erratic behaviour of robots properly, taking into account the "intentions" of such machines and moreover, by referring them to existing conventions of business and civil law. As stressed in Sect. 4.3.1, we should take the idea seriously that robots have intentions relevant in the civil (as opposed to the criminal) law, for intelligence emerges from the rules of the contractual game, rather than individual choices of the robotic agent. In the phrasing of Giovanni Sartor:

> [T]his leads to assimilate the situation of the user of [a robot] to the situation of a person handing over the conclusion of a contract to a human agent… What the two situations have in common, which distinguishes them from the situation where one uses a (mechanical or human) means of transmission, is cognitive delegation, *i.e.*, the decision to entrust the formation of the content of a contract and the decision whether to conclude it or not… to someone (or something) else's cognition (Sartor 2009: 280–281).

Admittedly, the current rules of legal systems bar the acceptance of the robots-as-agents approach in certain cases. Furthermore, there are key differences as to how common and civil law systems may aim to govern such technological applications. For example, in France or Italy, the legal personality of the agent is a necessary (yet not sufficient) requirement for acknowledging that machines can be proper agents in the civil (as opposed to the criminal) law field. *Vice versa*, in Anglo-American law, there is no objection "to the possibility of a nonperson artificial agent, on the grounds of a lack of capacity to contract in its own right on the part of the agent" (Chopra and White 2011: 56). Likewise, in the US, the principal is not bound by a contract that is outside the agent's actual or apparent authority, although a "minimum of physical and mental ability" or "volition" of the agent is required. In most civil (as opposed to common) law systems, the agent must be of sound mind, so that the risk of malfunction errors would fall on the third parties in all cases. However, despite this general disagreement, we should not overlook a crucial point: robots should be conceived as new proper agents in the civil law field because this legal option allows us to strike a fair balance between the individual's claim to not be ruined by the decisions of their robots and the claim of a robot's counterparty to be protected when doing business with them. Some brief remarks on the history of the law help us in the next Section: Roman lawyers addressed both legal agency of non-humans and guarantees for the counterparties interacting with them more than 2,000 years ago. A historical reference on the rules that governed the actions of slaves sheds light on how we could deal with today's robots following the pragmatic spirit of Roman law. The analysis of the ethical issues raised by this parallel, is postponed until Sect. 6.1.

4.4 Modern Robots, Ancient Slaves

The parallel between today's robots and slaves in ancient Rome seems appropriate, because slaves were considered as things that nevertheless played a crucial role in trade and commerce. In *The Human Use of Human Beings* (1950), the father of cybernetics, Norbert Wiener, suggested that "the automatic machine, whatever we may think of any feelings it may have or may not have, is the precise equivalent of slave labor." This similarity has been stressed time and again over the past years. In *The Responsibility of Intelligent Artifacts* (1992), Leon Wein reckons that automation is "bringing the conception of slavery back on the scene… As employees who replaced slaves are themselves replaced by mechanical 'slaves,' the 'employer' of a computerized system may once again be held liable for injury caused by his property in the same way that she would have if the damage had been caused by a human slave" (*op. cit.*, 111).

From a legal viewpoint, however, we should not miss the forms of agency that ancient Roman law admitted for such "things." Although most slaves certainly had no rights to claim against their own masters, some of them enjoyed a significant autonomy. The elite of the slaves, as in the case of the emperor's slaves, were estate managers, bankers and merchants, holding important jobs as public servants, or entering into binding contracts, managing and making use of property for their masters' family business. Consider the case of the *institor* (*Dig*. XIV, 3, 11, 3; XV, 1, 47). Such slaves managed different classes of convenience stores, *taverna*, such as bakeries and barbershops; wineries, hot drinks, or ready-prepared meat; and even, so to speak, booksellers' minimarts. When Emperor Nero was convinced to participate in the Olympic games of 67 A.D. in order to improve relations with Greece, it was not a joke that he entrusted his freedman Helios with the right to convict or seize anyone in Rome.

The parallel between robots and slaves is hence attractive, because the rules of ancient Roman law on slavery show a way to address certain of the inconsistencies of the robots-as-tools approach mentioned in the previous section. While Roman lawyers invented forms of agency and autonomy for mere things without legal personality, their aim was to strike a balance between the interest of the masters not to be negatively affected by the business of their slaves and the claim of the slaves' counterparties to be able to safely interact or do business with them. Today's idea that (certain types of) robots may be held directly accountable for their own behaviour has thus a precedent in the ancient Roman legal mechanism of *peculium*. In order to avert any legislation preventing the use of robots due to excessive burdens on the owners (rather than producers and designers) of these machines, the idea is that, at times, only "robots shall pay" could be the right answer.

4.4.1 The Digital Peculium

There is a key difference between criminal and civil lawyers dealing with new types of responsibility for the behaviour of robots. The focus of criminal lawyers is most of the time on harm or damages caused by such machines: something had to go wrong, in other words, so as to determine whether we are dealing with crimes of intent, negligence, or further legal observables examined with the phenomenology of *Picciotto Roboto* in the previous Chapter. *Vice versa*, it is not necessary that something has to go wrong in civil law: on the contrary, since the late nineteenth century, the legal imagination has been fired by how machines can be extremely fruitful in making contracts, or establishing rights and obligations between humans, in a win-win scenario. Although today's debate on cognitive automata in the form of software agents can be traced back to the seminal remarks of German scholars on automation and the law in the late 1800s, what technology has challenged over the past decades is the traditional viewpoint that robots are mere tools, rather than proper agents, in the legal field. Some reckon that we should register such machines just like corporations. This idea, for example, has been proposed by Curtis Karnow in *Liability for Distributed Artificial Intelligence* (1996), Jean-François Lerouge in *The Use of Electronic Agents* (2000) and Emily Weitzenboeck in *Electronic Agents and the Formation of Contracts* (2001). Certain scholars, as Anthony Bellia in *Contracting with Electronic Agents* (2001), suggest that we should bestow robots with capital. Others, as Giovanni Sartor in *Cognitive Automata and the Law* (2009), think that making the financial position of such machines transparent is a priority. Whilst further policies are feasible and even indispensable, *e.g.*, insurance models, what these proposals have in common has a precedent in the ancient Roman legal mechanism of *peculium*. According to the Digest of Justinian, the *peculium* was "the sum of money or property granted by the head of the household to a slave or son-in-power. Although considered for certain purposes as a separate unit and so allowing a business run by slaves to be used almost as a limited company, it remained technically the property of the head of the household" (Watson 1988: xxxv–xxxvi).

As a sort of proto-limited liability company, the *peculium* aimed to strike a balance between the claim of the masters not to be dilapidated by their slaves' businesses and commercial activities and the interest of the slaves' counterparties to safely transact with them. Most of the time, a master's liability was limited to the value of their slave's *peculium* and yet, the legal security of the latter guaranteed that obligations would have been met. For example, the contractual counterparties of the slaves could check whether the negotiations fell outside the authority or financial autonomy of the slave and, *vice versa*, in the wording of the Digest, "anyone who does not wish contracts to be made with him may prohibit it" by giving public notice (*Dig.* XIV, 3, 11, 3). Similarly, the mechanism applied when "the party desired business to be transacted with him under a certain condition, or through the intervention of a certain person, or under a pledge" (*Dig.* XIV, 3, 11, 5). But, going back to the case of the *institor* managing different classes of convenience stores, what did giving public notice mean?

> To give public notice we understand to mean that it shall be made in plain letters, so as to be easily read from the ground; that is to say, in front of the shop or place where the business is carried on, not in a retired place, but in one which is conspicuous. Shall the notice be in Greek or in Latin letters? I am of the opinion that this depends upon the character of the place, so that no one can plead ignorance of the letters…
>
> It is essential that the notice should be permanently posted; for if the contract was made before the notice was set up, or it was concealed, the Institorian Action will be available. Hence, if the owner of merchandise posted a notice, but someone removed it, or through age, rain, or something of this kind, the result was that there was no notice, or it did not appear; it must be said that the party who made the appointment will be liable. If, however, the agent himself removed it for the

purpose of deceiving me, his malicious act should prejudice the party who appointed him, unless he who made the contract also participated in the fraud (*Dig.* XIV, 3, 11, 3–4. Trans. by S. P. Scott, *The Civil Law*, IV, Cincinnati, 1932).

Matters of legal certainty, financial and contractual warranty, or transparency, can obviously be improved in the case of modern autonomous robots. When following the example of ancient Roman lawyers, however, we should distinguish different kinds of robo-traders, as Romans did with multiple types of activities and status of the slaves as *dispensatores*, *ordinarii*, etc., for each of which specific lawsuits or *actiones* were established: besides the aforementioned *Institorian* action, think about the *actio exercitoria*, *tributaria*, etc.[3] Therefore, we have to distinguish the kind of business or commercial activity the robot is entitled to pursue, whether the robot acts on its masters' behalf or as a mediator between third parties, while being understood that the behaviour of the robot will be bound by rules and conventions that usually apply to the circumstances of the case. Consider the (not too futuristic) case of a robotic personal assistant such as a sort of i-Jeeves that helps us schedule a set of conferences, lectures and meetings at several European (or US) universities. Whereas we may guess at the best way of accepting simultaneous invitations from Oxford, Barcelona, Heidelberg, Athens and Paris, our robot needs not resolve the travelling professor problem by determining the shortest possible tour that visits each university only once. Rather, we expect that i-Jeeves checks both the availability and convenience of logistics in accordance with a number of parameters such as budget, time efficiency, or weather average conditions: i-Jeeves reports its findings back for a decision or, even, could determine the steps of our tour by directly booking hotel rooms, flights and so forth. Such contracts would not only be valid but, thanks to the digital *peculium*, a fair balance would be struck between the different human interests involved. By employing robots or artificial agents to do business, transactions or contracts, individuals could claim a liability limited to the value of their robots' portfolio (plus, eventually, forms of compulsory insurance), while the robots' *peculium* would guarantee their human counterparties, or other robots, that obligations would really be met.

On the other hand, we can further the Roman legal framework by granting robots legal accountability. As occurs with traditional artificial persons, as seen above in Sect. 2.3.2, legal systems may sever the responsibility of designers, manufacturers, operators and users of robots dealing with third parties, so that, on the basis of the warranty of their own *peculium*, only robots would be held liable for damages caused by them. Admittedly, this solution has several advantages: on the side of the contractual counterparties

[3]For a more complete list, see Ŝtaerman and Trofimova (1975: 82).

of robots, the personal accountability of such machines renders irrelevant whether they are acting beyond certain legal powers and who should be held liable for conferring such legal powers. On the side of users and operators, the personal accountability of robots allows humans to evade responsibility for possible malfunctions of the machine as well as errors of induction and specification, as seen above in Sect. 3.3.2. Moreover, aside from the quantification of the *peculium* and data on which insurance policies might hinge, the personal accountability of robots seems to be particularly recommended for certain applications. In light of a new generation of AI chauffeurs and intelligent car sharing, let me examine this hypothetical separately in the last section of this chapter.

4.5 The UV Revolution

One of the most dynamic fields of robotics technology today deals with the design, production and use of Unmanned Vehicles ("UV"). Although the technology is currently more prominent in the military than the civilian sector, a number of factors such as inter-agency transfers, increasing international demand, public R&D support and growing access to powerful software and hardware, explain why the civilian use of this technology is rapidly and progressively mounting. This is the case for several UV applications such as for border security, law enforcement, emergency and hazard management, remote exploration works and repair, urban transport, farming and more. As Brendan Gogarty and Meredith Hagger argue in *The Laws of Man over Vehicles Unmanned* (2008), the relative cost savings promised by UV technology have "excited many commercial operators" (*op. cit.*, 110), so that it is crucial for lawyers to assess the regulatory constraints for the ever-growing production and use of this new generation of UVs. More particularly, attention should be paid to three types of unmanned vehicles.[4]

---

[4]As mentioned in Sect. 3.5, we should grasp the unmanned vehicles as part of a more complex multi-agent system where such autonomous or semi-autonomous machines interact with maintenance and safety contractors, traffic operators or internet controllers, in order to avoid communication interferences, environment concerns, collisions, and the like. By considering that such machines will increasingly be connected to a networked repository on the internet that allows robots to share the information required for object recognition, navigation and task completion in the real world, some scholars refer to this type of robots as intelligent unmanned systems, unmanned aircraft or rotorcraft systems, and so forth. The aim of this section, however, is to stress the different ways UAVs, UUVs, and UGVs may affect current legal frameworks, rather than the systemic features of such network-centric applications.

The first type is provided by aerial applications, that is, UAVs. As previously stated above in Sect. 3.3, more than forty countries are currently developing such a kind of technology for military purposes. In addition, there already are cases of non-lethal engagement of suspects, arrests by drones, monitoring operations and UAVs specifically designed for policing, patrolling and inspection. As Peter Singer stresses in *A World of Killer Apps* (2011), "police departments in cities such as Miami, Florida and Ogden, Utah, have sought special licenses to operate unmanned aerial surveillance systems." However, the advancement is so rapid that drones already are within the reach of public bodies, private companies and even individuals. Both the US and EU are adopting regulations and procedures so as to permit UAVs to share the same airspace as commercial traffic. Aside from the law enforcement field, consider the definition of aircraft and related products as contained in Article 3 of the European Regulation EC 216/08, which appears broad enough to include UAVs. Likewise, in the spring of 2011, the US Congress established that "US civilian airspace should be opened to allow more widespread use of such systems by 2015" (Singer 2011). Rather than issues of military immunity and criminal accountability as previously mentioned in Sect. 3.5, the civilian use of UV technology puts forward problems of human responsibility and contractual liability concerning safety claims such as control loss, link issues, automated recovery or piloting regulation.

The second type of UV technology is offered by water-surface and underwater ("UUV") applications such as in remote exploration work and repairs of pipelines, oil rigs and so on. Among UV devices, this is one of the most developed fields: Gogarty and Hagger have even spoken of the golden age of UUV technology that "occurred more than a decade before the UAV revolution" (*op. cit.*, 104). Whilst development in UUVs and the increase of their use in the civil sector are likely to force lawmakers to amend many clauses of the current legal framework in maritime law, *e.g.*, the 1972 IMO COLREGs Convention, it nonetheless seems that UUVs do not really affect basic tenets of the law. In light of today's spectrum of robotics applications, as seen above in Sect. 4.1, UUVs are in fact closer to reasonable safety and controllable machines such as the da Vinci surgery system, than the ultra-hazardous activity of (certain types of) UAVs. Although there are UUVs that autonomously undertake their work by preventing damage, alerting controllers or repairing oil rigs in the Caribbean Sea, the legitimacy of such automatic devices can be grasped by lawyers using the same concepts developed for previous technological innovations, that is, in terms of the probability of events and the cost of their consequences.

The third type of UVs finally offers some of the most challenging applications of this technology, namely, the civilian (rather than military) use of unmanned ground vehicles. Whether or not future UGVs will need driving licenses, special licenses, etc., UV cars and AI chauffeurs allow us to deepen the legal issues that are raised by the civilian use of both UAVs and UUVs. The complexity of the environment that designers and producers have to address increases the uncertainty and unpredictability of UGVs automatically driving on the freeways. As a matter of risk, these UVs are more similar to unmanned flying vehicles than unmanned ships exploring the deep ocean floor. Yet, contrary to the use of UAVs patrolling the air for law enforcement purposes, the risks of employing UV cars mostly regard contractual obligations and problems related to strict liability in the field of torts, rather than constitutional safeguards and human rights law. On this basis, proponents of UGV technology ask for "a major review and clarification of existing civilian traffic safety regimes and even the creation of a specific regulatory system for UVs" (Gogarty and Hagger 2008: 121).

The next section dwells on whether new forms of accountability for the behaviour of these machines, such as the digital *peculium*, fit the new generation of AI chauffeurs and intelligent cars. Then, in the final Sect. 4.5.2 of this Chapter, the focus is on how UGVs suggest that lawyers will increasingly address (or be pressed by) cases of extra-contractual responsibility, *e.g.*, robots damaging third parties rather than affecting their contractual counterparties. This scenario proposes a further type of responsibility, such as the *Aquilian* protection in Roman law.

4.5.1 AI Chauffeurs and Intelligent Car Sharing

Intelligent vehicles driving themselves on highways are a popular subject of Sci-Fi movies: over the past 50 years, however, a number of states, organizations and private companies have made the dream come true. In the 1960s, the idea of building fully autonomous UGVs has been seriously pursued in several countries such as the US, Japan, Germany and Italy. Two decades later, the European Commission began funding a project on autonomous vehicles, the Eureka Prometheus Project (1987–1995). In the late 1990s, the US Congress authorized the Defence Advanced Research Projects Agency ("DARPA") to organize a series of prize competitions for driverless cars in order to develop the military sector of UGVs and make one-third of ground military forces autonomous by 2015. Whereas there already is a panoply of US military UGVs such as TALON and Panther M-60 (see Singer 2009), the advancement of the civilian sector has been impressive.

Consider the aforementioned DARPA Grand Challenge competition. The first race was held on 13 March 2004, in the Mojave Desert, but none of the cars completed it. Just a year and a one-half later, five vehicles successfully finished the second race. Starting a rivalry such as the competition between Oxford and Cambridge in the annual boat race, the 2004 winner, *i.e.*, the Carnegie Mellon University's Red Team was defeated by the Stanford University's Racing Team on 8 October 2005. Two years later, Carnegie Mellon had the opportunity to take the revenge at the "Urban Challenge." On 3 November 2007, the third DARPA competition concerned a 96 km urban area race, to be completed in accordance with all traffic regulations and within 6 h. Due to the rapid advancement of technology, the challenge was not only to complete such a tortuous route, but to complete it as soon as possible. Teaming with General Motors in the Tartan Racing, Carnegie Mellon overtook the Stanford-Volkswagen car, taking 4 h 10 min and 20 s, at 22.53 km per hour, to cross the finish line first...

Three years later, in 2010, the European Commission promoted the "Intelligent Car initiative." As the corresponding website is keen to inform, the aim is to "imagine a world where cars don't crash, where congestion is drastically reduced and where your car is energy efficient and pollutes less." There are around 1.3 million mishaps and 41,000 people who die in car accidents on EU roads each year (whereas, in the US, more than 37,000 fatalities occurred in 2008). Besides, traffic jams impact on 10 % of the European major road networks and costs are estimated 50 billion per year, that is 0.5 % of EU GDP. Moreover, road transport accounts for more than one-quarter of the EU's total energy consumption. Therefore, in the phrasing of the Commission, "the Intelligent Car initiative is an attempt to move towards a new paradigm, one where cars don't crash anymore and traffic congestion is drastically reduced. Part of the i2010 strategy to boost Europe's digital economy, the Intelligent Car initiative is an answer to the need of citizens, industry and the Member States to find common European solutions and to improve the take-up of intelligent systems based on information and communication technologies ("ICT")."

Meanwhile, under the supervision of Sebastian Thrun, the director of the Stanford AI Laboratory and team chief of the robotic vehicle Stanley – which won the 2005 DARPA competition mentioned above – Google has been developing and testing its own driverless cars. As of 2010, such vehicles have driven 230,000 km with some human intervention and 1,600 km completely alone. A year later, lobbied by Google, the Nevada Governor signed into law a bill that, for the first time ever, authorizes the use of autonomous vehicles on public roads. Approved by the Nevada Assembly (36–6) and the Senate (20–1), the law amends certain provisions governing transportation and, furthermore, establishes that the Nevada Department of Motor Vehicles "shall adopt regulations authorizing the operation of autonomous vehicles on highways within the State of Nevada" (AB 511, June 2011). Although such regulations on safety and performances standards may take a long time, what is at stake here concerns experimental cars where "a human driver can override any error," as John Markoff reports in The New York Times, quoting some Google researchers.[5]

Still, it is a short step to envisage fully autonomous UGVs driving themselves in Nevada and, for that matter, spreading ubiquitously on public roads. However, despite rapid advancement of technology in key components of such cars as adaptive headlamps and cruise control, blind spot monitoring and driver checking systems, traffic sign recognition, pre-crash schemes and so forth, it is likely that lawyers should be prepared to address a new class of hard cases. In fact, who should be liable if the autonomous car has an accident? In the phrasing of *The Laws of Man over Vehicles Unmanned*, how will fault be determined when a human and computer are sharing the reigns of a vehicle under traffic legislation? Indeed, who will be at fault if the vehicle has an accident when it is clear only the computer AI was in control? (Gogarty and Hagger 2008: 120–121). Moreover, in the name of urban sustainability and green policies stressed above, how about new forms of distributed responsibility as soon as we reflect on, say, schemes of AI car sharing?

As mentioned above in Sect. 4.3.2, traditional forms of apportioning individual liability fall short in coping with such scenarios. Let me insist on three points:

First, there is the difficulty for traditional legal outlooks of addressing the behaviour of robots as agents, rather than simple instruments of human interaction. As a matter of fact, humans will delegate to such autonomous and even intelligent cars complex cognitive tasks, such as driving themselves on the highways, while avoiding other cars, preventing individuals' reckless manoeuvres and so forth.

Second, from the fact that a human let the car drive by itself, it does not follow that the legal effects of the decisions of that car should necessarily fall upon the human. On the one hand, we are back to cases of apportioned responsibility of designers, manufacturers, dealers and users of AI machines, which inspired Curtis Karnow to predict a failure of legal causation as discussed above in Sect. 3.5. On the other hand, the hypothetical of environmentally friendly-AI car sharing makes this scenario still more complex, since such machines would be dealing with a multitude of human masters.

Finally, we should take into account the protection of third parties. Compared to the form of agency in the case of robo-traders, the spectrum of third parties widens so as to transcend the field of contractual obligations and concern what common lawyers call torts, that is, in the jargon of civil

[5] *Google Cars Drive Themselves, in Traffic*, October 10, 2010, A1 of the New York edition.

lawyers, forms of extra-contractual liability. In the case of robo-traders, individuals grant them authority to act on their behalf when dealing with third parties, so as to accept bids, make offers, compare prices, etc. In the case of AI chauffeurs, individuals will grant them authority to autonomously drive on the freeways, so that, theoretically speaking, everybody could be affected by the reckless behaviour of these machines.

A new form of accountability, such as the digital *peculium*, that could successfully tackle the legal challenges of a new generation of UGVs was introduced in Sect. 4.4.1. After all, we can imagine AI chauffeurs that accept offers, or make contracts, so as to autonomously drive individuals on the streets. Therefore, on the side of the contractual counterparties of robots, the personal accountability of AI chauffeurs guarantee that obligations for damages caused by such machines would be met. On the side of both users and operators, the personal accountability of AI chauffeurs let people evade liability for possible unpredictable malfunctions of the machine. Whilst it is crucial to determine the sum of money granted to the intelligent car, it is likely that programs such as Google's driverless cars or the European Commission's i2010 strategy will provide enough data on the probability of events, their consequences and costs, to determine levels of risk and, therefore, both the amount of the *peculium* and forms of compulsory insurance, on which new forms of accountability for the behaviour of such machines may hinge. This is the approach suggested by a number of scholars, such as Tom Allen and Robin Widdison in *Can Computers Make Contracts?* (1996), Ian Kerr in *Ensuring the Success of Contract Formation in Agent-Mediated Electronic Commerce* (2001), Woodrow Barfield in *Issues of Law for Software Agents* (2005), Francisco Andrade et al. in *Contracting Agents: Legal Personality and Representation* (2007), down to the aforementioned works of Giovanni Sartor (2009) and Chopra and White (2011).

However, would new forms of personal accountability for robots represent the one-size-fits-all answer to the new generation of legal issues brought on by such robots? Does this approach apply equally to robots as agents and robots as instruments? Does the legal accountability of the robot suffice to deal with different types of claims in the field of torts?

4.5.2 Unjust Damages

We have examined three different types of robots in this Chapter. First, we dwelt on robots as means of human industry and interaction that include both ends of the spectrum of robotic applications as examined in Sect. 4.1; namely, reasonable safe and controllable machines, such as the da Vinci surgery system, and the ultra-hazardous activities performed through some of today's UAVs. As means of human industry, such machines do not challenge basic tenets of the law as current provisions of contracts and tort law properly address damages or harm caused by these robots. Think of strict product and malfunction liability claims, breach of warranty, negligence, or evidence, that is, the set of concepts examined through the mechanism of the burden of proofs in the *Mracek v. Bryn Mawr Hospital* case discussed above in Sect. 4.2.2. As Richard Posner affirms in *Economic Analysis of Law* (1973), "new activities tend to be dangerous because there is little experience with coping with whatever dangers they present... The fact that the activities are new implies that there are good substitutes for them" (*op. cit.*, 2007 edition: 180).

A second class of robotics applications has to do with robots as legal agents. Rather than simple objects concerning clauses and conditions of contracts, the example of certain robo-traders has shown machines capable of determining clauses and conditions of contracts by themselves. Here, current provisions of the civil (as opposed to the criminal) law fall short in addressing both the cognitive states of such machines and ways for determining or apportioning liability for damages caused by this class of robots. Some ways for severing the chain of responsibilities between designers, manufacturers, operators, users and third parties that interact with such machines, were discussed above in Sect. 4.2.2 and Table 4.1, according to three different kinds of erratic behaviour: robotic specification, induction, and malfunction of the robot. Whereas traditional legal standpoints end up in a Hegelian night, where all kinds of liability are blurred into the same grey colouring, we should define where to cut back on the scale of the activity. New forms of accountability for robots as strict agents in the civil law field, *e.g.*, the digital *peculium*, show how to prevent this threat, so as to "cope with whatever dangers they present" (Posner 2007). By granting authority to the robot, so as to let it act on an individual's behalf when dealing with third parties, a new form of *peculium* strikes a fair balance between the counterparties of robots demanding the ability to safely interact or transact with such machines and individuals claiming that they should not be ruined by the decisions or behaviour of their own robots. Although it would be meaningless to treat the first class of robots, *i.e.*, robots as means as legal persons with a contracting capability in their own right, it makes a lot of sense to attribute such capability to the new generation of robo-traders.

Finally, there is the class of robots as intermediates in social life, rather than agents of human business and negotiations. As the example of the AI chauffeurs has shown, such robots can make business and still most of the time, they will be dealing with third parties, namely, individuals who are not directly concerned by the enforcement of rights and obligations created by the robots' business. In the phrasing of the UN 2005 Robotics Report, this class of machines concerns "domestic or personal use of service robots for domestic tasks, entertainment, handicap assistance, personal transportation, home security and surveillance." Such a class of robots as intermediates of human interaction brings us back to the scenario of AI chauffeurs provoking accidents on the highway. Consider a new generation of robot toys (entertainment), or robot nannies (domestic tasks and handicap assistance). In the case, say, a nanny such as Jetsons' Rosey, nursing your old mother, causes harm to some of your mother's acquaintances, who is liable?

This scenario goes beyond the contractual mechanism of *peculium* and involves what Roman jurists defined in terms of *Aquilian* protection; namely, the form of responsibility stemming from the general idea that individuals are held liable for unlawful or accidental damages caused to others because of their personal fault: *Alterum non laedere* as discussed above in Sect. 2.2. Although the digital *peculium* may govern certain cases of extra-contractual responsibility, *e.g.*, road accidents, there is a number of further obligations, so as to protect from unjust damage, in the many-to-many, rather than one-to-one contractual scenarios of social interaction. Think of strict liability rules in the field of robotics by analogy with dangerous animals as seen above in Sect. 3.4.3. Likewise, consider cases of liability for the negligent control of artificial agents and even vicarious responsibility for the autonomous acts of individuals' artificial employees. What is crucial here concerns the different robotic applications with which we are dealing, since such robots as domestic service robots, as a sort of AI children, animals, or i-Jeeves, entail different types of liability and opposite ways to determine on whom the burden of proof should fall. These are cases where we need a further type of expertise in the laws of robots. After the chapters on crimes and contracts, we will deepen the examination of that which common lawyers define as the field of torts.

Chapter 5
Torts

If we wait for the moment when everything, absolutely everything is ready, we shall never begin

Ivan Turgenev, Fathers and Sons

Abstract Attention is drawn to issues of extra-contractual responsibility, *i.e.*, when robots damage third parties rather than their contractual counterparties. What common lawyers define as torts deals with obligations between private persons imposed by the government to compensate for damage done by wrongdoing. Here, the new class of hard cases that the growing autonomy of robots is likely to induce, concerns how we should interpret a novel kind of liability for the behaviour of others. For the first time ever, legal systems will hold humans responsible for what an artificial state-transition system "decides" to do. Moreover, this kind of liability crucially depends on the different kinds of robots with which we are dealing: a robot nanny, a robot toy, a robot chauffeur, a robot employee, and so forth. This is one of the most innovative aspects in the field of the laws of robots, as traditional forms of responsibility for the behaviour of others, such as children, pets, or employees, have to be complemented with new strict liability policies, or alternatively, mitigated through insurance models, authentication systems, and the mechanism of allocating the burden of proof.

There is a further set of cases involving individual responsibility beyond that of criminal and contractual liability. These cases arise based on damages caused to others because of personal fault. This kind of extra-contractual responsibility, which common lawyers define as torts, was at stake in *Mracek v. Bryn Mawr Hospital* as discussed above in Sect. 4.2.2. The plaintiff's

U. Pagallo, *The Laws of Robots: Crimes, Contracts, and Torts*, Law, Governance and Technology Series 10, DOI 10.1007/978-94-007-6564-1_5,

claims in fact revolved around damages arising out of strict product and malfunction liability, alleging that designers and producers of robots should be held liable for damages caused to third parties by the defective manufacture of the product or flaws in the design. The difference between such forms of liability can be appreciated with the mechanism of the burden of proof. In the U.S., for instance, plaintiffs have to demonstrate the defectiveness of the product under the manufacturer's control as the proximate cause of the injuries suffered in cases of strict product liability. *Vice versa*, dealing with strict malfunction liability, direct evidence as to the defective condition of the product or the precise nature of the product's defect is not necessary. Rather, plaintiffs have to prove the existence of that defect through the circumstantial evidence of the occurrence of the malfunction, or through evidence eliminating any abnormal use of the product as well as reasonable secondary causes for the accident. This complex set of notions and ways of determining on whom the burden of proof falls, gives rise to the extremely detailed, and sometimes strange, labels on products whereby which manufacturers warn about risks or dangers involving the improper use of the artefact, *e.g.*, a robot. Whilst an imposition of strict liability often can depend on the provision of inadequate warnings, or lack of information about certain features of the product, we may speculate about the rationale for this type of tort liability. According to Posner's *Economic Analysis of Law*:

> The economic rationale for strict product liability is that consumers can do little, at reasonable cost, to prevent a rare product failure. Imposing the costs of accidents on the manufacturer will lead to price increases, resulting in consumers substituting toward other, less dangerous, products. The activity consisting of the manufacture and sale of less safe products will diminish and with it the number of product accidents. Strict liability effectively impounds information about product hazards into the price of the product, causing a substitution away from hazardous products by consumers who may be completely unaware of the hazards (Posner 2002: § 6.6).

Along with strict liability rules, two types of negligence-based responsibility were also considered in the previous chapters of this book. On one hand, the third (and final) step of the phenomenology of *Picciotto Roboto* has to do with cases of liability for the behaviour of robots based on the negligent control of the artificial agent as seen above in Sect. 3.4.3. In that context, the example was a robot attacking some friends during a garden party at my villa: here, the example can be adapted to the field of tort law, imagining the robot owned by a friend breaking my wife's sixteenth century Delftware vase during the same party. On the other hand, plaintiff's claims in *Mracek v. Bryn Mawr Hospital* concerned not only damages arising out of strict product and malfunction liability, but also a negligence-based liability of designers and producers of robots having a duty to conform to a certain standard of conduct. Mracek claimed in fact that his counterparties breached that duty, thereby provoking an injury and actual loss to the plaintiff. In such cases, individual liability is based on a lack of due care, namely, the duty of the reasonable person to guard against foreseeable harm. When the robot is, say, an ISO 8373 industrial robot, traditional cases of negligence-based liability follow as a result, as already examined in Sects. 3.4.3 and 4.3.2. Yet, if the robot is a service machine for domestic or personal use, there are three reasons why lawyers will probably have to address an increasing number of hard cases.

First, envision the plaintiff's burden of proof for negligence-based product liability and the capability of robots to gain skills from their own interaction with the environment and humans as caretakers of such robots. The more these machines are adaptable, interactive and autonomous, the more users will find it difficult to prove that the manufacturer of the robot did not conform to a certain standard of conduct, or that the supplier did not guard against foreseeable harm. Does this scenario challenge the economic rationale for today's strict liability rules?

Second, negligence-based liability for the use of such robots will most likely be added to the current strict liability safeguards in the field of tort law. This scenario is not new, since it traditionally applies to individuals' responsibility for the behaviour of their animals, employees and in most legal systems, their own children. However, it is far from clear how legal systems will tackle cases of negligence-based liability for the use of these domestic robots. Should they be likened to the current rules of strict liability for the behaviour of animals, children or employees?

Third, according to certain scholars, we should be ready to tackle a new generation of intentional torts, such as liability for wrongful conduct that can be ascribed to a human because, for example, her service robot "aimed" to do harm.[1] Here, we do not have to buy the idea that robots may have human-like intentions in order to admit a new generation of cases concerning responsibility for the behaviour of others depending on how individuals treat, or take care of, their own robots.

Contrary to the fields of criminal and contract law, we do not have clear-cut canons of tort law, such as the principle of legality in criminal law, or the autonomy of the contractual parties and their agreements in civil law, through which to define most of the new cases of tort liability. Admittedly, the production and employment of service robots for personal and domestic use are still in their infancy and yet, no divinatory powers are needed to

[1]This is the viewpoint of the front of robotic liberation as seen above in Sects. 2.1.1 and 3.1.

expect a dramatic increase in their use within the next few years. Therefore, I would admit that "the owl of Minerva takes its flight only when the shades of night are gathering" and still, *pace* Hegel and his warning in the *Philosophy of Right*, we have to explore the set of principles, concepts and ways of legal reasoning that probably will be affected by tomorrow's service robots, domestic AI machines, and so forth.

Next, the focus in Sect. 5.1 is on a first kind of extra-contractual responsibility for the use and even the design and construction of such robots, that is, cases of "intentional" torts. Whilst this scenario is closely related to the phenomenology of *Picciotto Roboto*, special attention is drawn to the thesis of Richard Posner that "the concept of intent is merely a stopgap." Although for different reasons, the aim of this section is to show why the notion of "intent," *pace* the front of robotic liberation, is not crucial in the field of robotic torts.

Section 5.2 deals with a second kind of tort liability based on lack of due care. In order to understand how responsibility is established in cases of negligence, the focus is on how the burden of proof works. For example, in some legal systems, parents evade responsibility when they can prove that they could not prevent their child's behaviour. Likewise, owners of animals are not liable if they can prove that a fortuitous event happened. Whether or not (some types of) robots should be considered as a sort of AI minor or conversely, a smart pet, the main legal issue with respect to the new generation of robots for personal or domestic use will often concern how we train, treat or manage our machines, rather than around who owns, builds or sells them.

The final type of tort liability is treated more exhaustively in Sect. 5.3, namely, the responsibility that the law imposes regardless of the person's intention or use of ordinary care, as occurs with employers' responsibility for the behaviour of their employees. As a form of distributing risk and responsibility, most legal systems establish that employers are strictly liable for any harmed caused by the actions of an employee engaged in during her work contract activities. This form of vicarious responsibility illustrates the current state-of-art in the laws of robots, according to which responsibility for damages caused by domestic and personal robots should be determined on the basis of the current strict liability rules governing cases of responsibility for the behaviour of employees. The aim of this section is to explore how this strict liability regime can be mitigated, so as to promote (and protect humans against) the use of service robots for personal and domestic use.

Finally, matters of tort policy are examined in Sect. 5.4 in connection with today's debate on the precautionary principle and how, in the name of precaution, the burden of proof should shift from those suspecting a risk in the construction and use of robots, to those who discount that risk. By determining who has to prove what, in accordance with the type of personal robot or domestic machine with which we are dealing, this approach lays the groundwork for the final chapter of this book concerning the law as a meta-technology.

5.1 Bad Intentions

Extra-contractual obligations, generally imposed against the will of the party seen as invoking in some sense the harm, can be distinguished into three categories: intentional torts, negligence-based responsibility and strict liability (Gordley 2006). Contrary to that occurring in criminal law with the principle of legality, clauses and provisions of tort liability are "open," that is, courts may determine the unlawfulness of certain behaviours by drawing parallels with previous cases. While technological innovation forces lawmakers to intervene by adding norms to the regulation of new (circumstances of new) crimes, courts can define matters of robotic tort liability, notwithstanding the novelty of such cases, in accordance with principles of decision inferred through analogy with precedents of tort law. This is not to say, of course, that questions of legal right and tort liability should be resolved by the exercise of simple discretion. Rather, matters of legal analogy suggest that we should ascertain whether the advancement of robotic technology affects the ways jurists have traditionally dealt with the field of torts. After the adventures of *Picciotto Roboto* in criminal law, should we sketch a phenomenology of robotic torts? How about the class of torts hinging on the voluntary wrongdoing of the tortfeasor?

Remarkably, several scholars have strenuously criticized this very idea of "intent." For example, in *The Jurisprudence of Skepticism* (1988), Richard Posner argues that "the notion of 'intent' plays no role other than as a proxy for certain characteristics of the tortious act, notably a big disparity between the cost (great) of the act to the victim and the (small and even negative) cost to the injurer of avoiding the act… It is a confession of ignorance, and if economics can help us to dispel the ignorance it may help us to dispense with the concept [of intent]" (*op. cit.*, 868). Furthermore, according to Posner, we should abandon the very idea of intention in criminal law. In the phrasing of *The Jurisprudence of Skepticism*, "the role of mental entities in law, such as 'intention,' should diminish as law becomes more sophisticated," because "as law matures, liability – even criminal liability – becomes progressively more 'external,' that is, more a matter of conduct that of intent" (*ibid*).

There are indeed cases where an individual's intentions should not be relevant. For example, when examining the just causes of war in Sect. 3.3.2, I stressed that military commanders and political authorities should be strictly responsible for all the decisions of robot soldiers in battle. Moreover, the laws of robots suggest further cases where we should follow Posner's ideas and dismiss the role that intentions play in determining tortious responsibility. Theoretically speaking, there are three such cases:

(a) The intentional tort that a human aims to carry out through her innocent robotic agent, but the machine deviates from the plan and commits some other offense;
(b) The intentional tort that a human perpetrates in cahoots with an evil robot; and
(c) The intentional tort that a robot commits notwithstanding its innocent human master.

Hypothesis (a) brings us back to the perpetration-by-another liability model in criminal law as seen above in Sect. 3.4.2. *Vice versa*, hypotheses (b) and (c) belong to the Sci-Fi pictures of morally wicked robots where the conduct of the machine, rather than the intention of any humans, is relevant. Hypothesis (b) is a futuristic example of the accomplice responsibility model in criminal law, as discussed in Sect. 3.4.3. Conversely, according to today's state-of-art, responsibility in hypothesis (c) would necessarily depend on human negligence.

However, most legal systems and scholars are reluctant to buy all of Posner's views. In fact, as stressed in the introduction to Chap. 3, the "intentional stance" often represents the only coherent strategy for describing and foreseeing the behaviour of complex entities, such as humans and some types of robots that can act in a teleological way. In addition, it is highly problematic to grasp the whole set of issues concerning tortious and even criminal liability as a matter of great vs. small costs, *e.g.*, cases of tort liability that overlap with the criminal accountability of the agent and the *mens rea* of humans. Moreover, the principles of equality and justice suggest that different cases should be treated in different ways, as occurs in criminal law, and how judges and juries address cases of, say, manslaughter, such as a cruel homicide or a heinous assassination, in order to quantify the punishment. This is not to say that such scenarios of "bad" intentions are particularly challenging in the laws of robots. In criminal law, humans who use their robots as instruments to commit some wrongful action are held strictly responsible, that is, they should pay their debt to society even though the robot deviated from the plan and carried out some other offence. In the law of contracts, the wrongful conduct of users of robotic applications would sever the link between claims of extra-contractual responsibility and previous contractual obligations. In tort law, the traditional legal viewpoint conceives either robots as dangerous animals or their use as an ultra-hazardous activity, whereas strict liability rules apply to all the circumstances. As a result, we can leave aside hypotheses of tort liability that depend on the evil intentions of humans, so as to focus on notions of reasonable foreseeability, vicarious responsibility and due care. From this stance, we can grasp the new generation of hard cases that will concern today's strict liability rules and forms of negligence-based liability for the use of robots. Whilst strict liability rules may fall short in coping with crimes of intent,[2] and the law of contracts,[3] it also is far from clear how lawyers should tackle cases of negligence-based liability for damages caused by robots employed for personal or domestic use, such as robot toys or robot nannies. In all these cases, work on human-robot interaction ("HRI") seems particularly relevant: By paying attention to different types of contacts with humans, robot functionalities and roles, as well as requirements of social skills, *e.g.*, the capability of robots to show aspects of human-style social intelligence, HRI approaches can help us to understand key features of the human-robot interaction that should be taken into account when examining tort liability for robotic behaviour.

The focus next is on the HRI work concerning the "caretaker paradigm," that is, humans as caretakers of robots. In the wording of Kerstin Dautenhahn's *Socially Intelligent Robots* (2007), attention should be drawn to the roles of humans that "identify and respond to the robot's emotional and social 'needs'. The human needs to keep the robot 'happy' which implies showing behaviours towards the robot characteristic of behaviour towards infants or baby animals." In light of this popular analogy in robotics, the aim is to examine how this parallel works in the laws of robots and more particularly, in the field of torts.

5.2 Children, Pets and Negligence

The expanding interactivity, adaptability and autonomy of robots have suggested in recent years a parallel with children and baby animals. In *Guilty Robots, Happy Dogs* (2008), David McFarland suggests that we are dealing with "alien minds" that force us "to take a further leap into the unknown," because we should teach robots to distinguish right from wrong much as we do with our children and pets. In *Moral Machines* (2009), Wendell Wallach and Colin Allen similarly stress the aim "to build machines that are capable

[2]See above in Sects. 3.4.2 and 3.5.
[3]See above in Sect. 4.4.2.

of telling right from wrong," so as to balance the goals and risks of the behaviour of robots and other artificial agents, and keep them within limits that individuals can accept. In legal terms, this responsibility primarily concerns designers and manufacturers, rather than the users of these machines. Accordingly, we examined in Sect. 3.4.1 the criminal features of this responsibility in connection with the first step of our phenomenology, namely, *Picciotto Roboto* by design. A spectrum of robotics applications was then illustrated in Sect. 4.1 in order to determine how the design and engineering of robotic applications can impact on clauses and conditions of contractual obligations. That which Frances Grodzinsky, Keith Miller and Marty Wolf (2008) present as the new "strong moral responsibilities" of designers and manufacturers of robots will be further examined below in Sect. 5.4. Nonetheless, robotic software and hardware programming are essential, but insufficient, conditions for establishing responsibility for the behaviour of these machines in the field of tort law.

Significantly, the annual IEEE RO-MAN series originating in Japan has focused since 1992 on the social behaviour, communication and intelligence in natural and artificial systems. Since robots are not a simple sort of "out of the box" machine, their behaviour may crucially depend on the ways individuals train, treat or manage them. When inspecting the NAO robot at the 2009 AISB conference in Edinburgh, I was impressed by how the Aldebaran's team had to teach the robot to use its own 57 cm tall humanoid body, let alone its on-board NAOqi software system, in order to move, walk, dance and interact with humans or other robots. At the 2010 AISB conference organized by the de Montfort University in Leicester, I could even appreciate NAO's improvement in being able to play its own violin! By following the viewpoint of current research on human-robot interaction, let us thus distinguish between a human-centred HRI approach and a robot-centred HRI methodology. In the first case, the idea is to keep robots within limits that people can rationally accept: in the words of *Socially Intelligent Robots*, "human-centred HRI is primarily concerned with how a robot can fulfil its task specification in a manner that is acceptable and comfortable to humans" (Dautenhahn 2007: 684). *Vice versa*, in the case of a robot-centred HRI approach, the emphasis is on the "robot as a creature, *i.e.*, an autonomous entity that is pursuing its own goals based on its motivations, drives and emotions" (*op. cit.*, 683).

This latter perspective seems particularly useful in order to understand how the legal responsibility of users, rather than the designers and manufacturers of robots, should be grasped in the field of torts and more particularly, in cases of negligence-based liability. Although the "social needs" of the robot are defined by the designer and modelled by the internal control architecture of the machine, it is the user that enables the robot to "survive in the environment" by fulfilling its needs. In *Designing Sociable Robots* (2002), Cynthia Breazeal's seminal work on Kismet, a robotic head with facial features, shows how this robot-centred methodology works. By treating the machine as an autonomous entity pursuing its own goals based on its motivations, humans indeed have to satisfy its social drives by singling out and responding to the robot's internal needs:

> The robot is treated as a "baby infant" or "puppy robot" with characteristic specific and exaggerated child-like features satisfying the "Kindchenschema" (baby pattern, baby scheme, schema "bebe"). The Kindchenschema is a combination of features that are characteristic of infants, babies or baby animals, which appeals to the nurturing instinct in people (and many other mammals) and trigger respective behaviours. The concept of the Kindchenschema goes back to the ethnologist Lorenz, who claimed that when confronted with a child, certain social behaviour patterns involved when "caring for the young" are released by an innate response to certain cues typically characterizing babies (Kerstin Dautenhahn, *Socially Intelligent Robots*, cit., 698, quoting Breazeal's research and Karl Lorenz's 1971 work on *Part and Parcel in Animal and Human Societies*).

It does not follow from a robot-centred HRI approach that humans must conceive such robots as though they were real pets or human cubs. For instance, in *Robotics Pets in the Lives of Preschool Children* (2006), Peter Kahn et al. examined children's interaction with the Sony's robotic dog AIBO in order to determine whether such interaction could blur foundational ontological categories and impact on children's social and moral developing. Although such artificial pets may induce protective feelings and even invoke a mutual double anticipation, Kahn et al. demonstrated that children do not perceive the AIBO as if it were a real dog and moreover, do not ascribe to it any moral standing.

However, it is not so difficult to imagine more complex cases, where social interaction with robots may involve emotional, physical and physiological activities that have a cost even for adult human beings. Whether humans will get the same payoff and gratification from their interaction with robots as they do with other human fellows is an open question that mostly depends on the cultural context and the type of application with which we are dealing: affective robots, sex tobots, carebots, medibots, AI chauffeurs and so forth. Some wonder if it is "ethically justifiable to aim to create robots that people bond with, *e.g.*, in the case of elderly people or people with special needs" (Dautenhahn 2007: 699). Others, like Peter Sullins in the introduction to *Open Questions in Roboethics* (2011: 236), provocatively affirm that, at least in the field of affective robots, "we might begin to prefer the company of machines." Furthermore, in *Love and Sex with Robots* (2007), David Levy argues that it is somehow inescapable that such machines will soon be widespread in our society, since this technology can fulfil many individuals' dreams and desires. Aside from the moral aspects of the debate, how should legal systems govern the widespread use of (some of such) robots? In particular, what about damages caused by this new generation of domestic robots that depend on the negligence of the human master?

By taking into account the parameters of the current research in human-robot interaction, it is likely that such negligence will more concern the ways individuals treat their robots, than how manufacturers design robots to fulfil their task specifications. Once "out of the package" the same model of robot will behave quite differently only after a few days or weeks, depending on how humans play their role of caretakers, so that the individuals' responsibility will hinge, at times, on whether they met the social drives of their own robots, detecting and responding to the robot's internal needs. On this basis, we can thus draw a fruitful analogy between traditional responsibility for the behaviour of others in tort law, *e.g.*, animals and children, and new scenarios of negligence-based liability for the conduct of robots. Rather than traditional responsibility for robots as means of human interaction, what is at stake with a new generation of robots for domestic and personal use concerns the duty of care that a reasonable person has to guard others against foreseeable harm. Next, the focus in Sect. 5.2.1 is on the regime of negligence-based liability in the U.S. field of torts. This is compared in Sect. 5.2.2 to a civil (as opposed to the common) law approach to the field of extra-contractual obligations, namely the Italian Civil Code. This latter perspective introduces the analysis of strict liability rules for damages caused by autonomous robots as found in Sect. 5.3.

5.2.1 American Parents

In *A Legal Theory of Autonomous Artificial Agents*, Chopra and White distinguish five types of negligent-based liability for individuals bearing responsibility for the care of other agents. Since "comparisons and analogies are provocative and serve to illustrate how the varied and enhanced abilities of artificial agents and the broadening range of responsibilities delegated to them will lead to comparisons with agents and other actors in diverse areas of law" (*op. cit.*, 135), Chopra and White propose the use of the traditional relations between principal and agent, master and servant, parent and child, warden and prisoner, as well as keeper and animal. In this context, we can leave aside parallels of robots with agents, servants and prisoners, so as to focus attention on the parallel between robots, children and pets. By filtering out the "legal variables" of Fig. 5.1, this stricter perspective suffices to allow us to understand how we can figure out the individual's negligent-based liability for the behaviour of (some types of) robots:

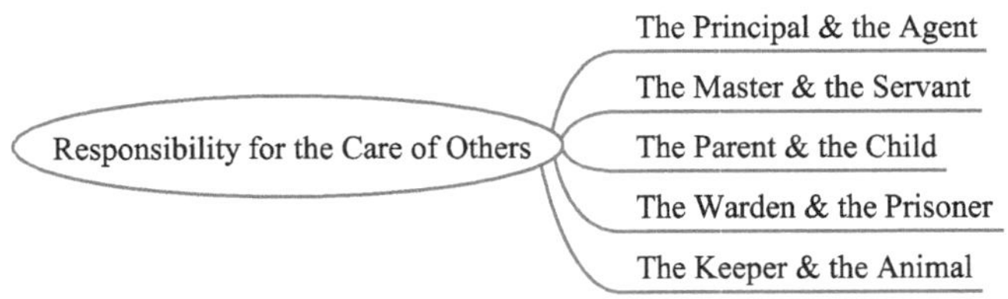

Fig. 5.1 A common law approach to negligence in the law of Torts

First, in the wording of Chopra and White (2011), "there might thus be analogies relevant to artificial agents in the duty placed on parents to take reasonable care to control minor children so as to prevent them from intentionally harming others or from creating an unreasonable risk of bodily harm to them" (*op. cit.*, 133). Contrary to most civil law systems, such a responsibility of American parents hinges on the scary personality of the minor and evidence that parents had knowledge or perception of this very fact. In the phrasing of Randall Hanson in *Parental Liability* (1989: 28), there is negligence-based liability for damages caused by children where "it can be shown that the minor had a propensity to cause a particular type of harm or injury and that the parents were aware of the dangerous propensity. If parents observe a recurring dangerous activity, they must take action to correct the child's activity or the parents may face liability on a negligence claim."

On the other hand, in the case of liability for harm caused by an individual's own animals, we should distinguish between animals that are known or presumed to be dangerous to mankind and domestic pets. The first hypothesis was considered in Sect. 3.4.3: owners or keepers of dangerous animals are held strictly liable for any damage caused by them, that is, regardless of any illicit or culpable behaviour of owners and keepers of such animals. *Vice versa*, when harm or damages are caused to third parties by an allegedly peaceful pet, U.S. tort law establishes a curious parallel with the responsibility of parents to take care of their minors. In the phrasing of Chopra and White, "a keeper of domestic animals is subject to negligence-based liability for injuries inflicted by her animals where the keeper has been negligent, the animals were wrongfully in the place where they inflicted the injuries, and the injuries are the result of known vicious tendencies or propensities" (*op. cit.*, 134). Yet, if "the keeper must have known or had reason to know of a dangerous propensity or trait that was not characteristic of a similar animal" (*op. cit.*, 130), such a keeper (or owner) will be held strictly liable for every injury caused by such problematic pet.

However, owners or users of robots in the foreseeable future will hardly be able to discern whether their machine presents any dangerous propensity or trait that is not typical of similar models. Moreover, the increasing autonomy and unpredictability of robotic behaviour will make it difficult for users or owners of such machines to evade responsibility, claiming that any injuries, damages or harm caused by their robots was reasonably unforeseeable. In addition, the capability of such machines to gain knowledge and skills from interaction with human caretakers, suggests that the fault would rarely fall on the designers, manufacturers or suppliers of such robots. Rather, according to the rationale for strict liability rules, it could be argued that owners or users of robots are in the best position to understand what is going on with the machine, so as to prevent its dangerous behaviour, regardless of whether the conduct of the robot was typical of similar robots, reasonably foreseeable and so forth. Whereas the risk is that individuals will think twice before purchasing and using robots for domestic services and personal fun, we may of course introduce insurance policies so as to avert this risk. Besides, in the long run, *i.e.*, after two or three generations of AI children or smart artificial pets interacting with their human caretakers, we can suspect that the duty of humans to take care of such machines will not be deemed similar to the current responsibility to control the dangerous propensities of animals and children. However, the question can be raised whether users and owners of robot toys and robot nannies need to wait for the long run in order to be finally reckoned as the reasonable person of today's tort law. Furthermore, would an analogy to today's liability of American parents be the only way to approach tomorrow's cases of negligence-based liability for the behaviour of service machines and domestic robots?

5.2.2 Italian Parents

So far, we have examined the field of torts in accordance with the Anglo-American partition concerning Posner's "stopgap" of intentional wrongdoing, negligence-based responsibility and strict liability. Admittedly, this is not the exclusive way of focusing on that which civil (as opposed to common) law scholars call extra-contractual responsibility. For example, Article 2043 of the Italian Civil Code adopts the principle of the Roman law tradition, namely, *alterum non laedere*, according to which individuals are liable for harms caused to others because of their personal fault as seen above in Sects. 2.2 and 4.5.2. On this basis, the Italian Civil Code determines two cases where individuals can evade such responsibility, namely, "self-defence" (Article 2044) and a "state of necessity" (Article 2045). The

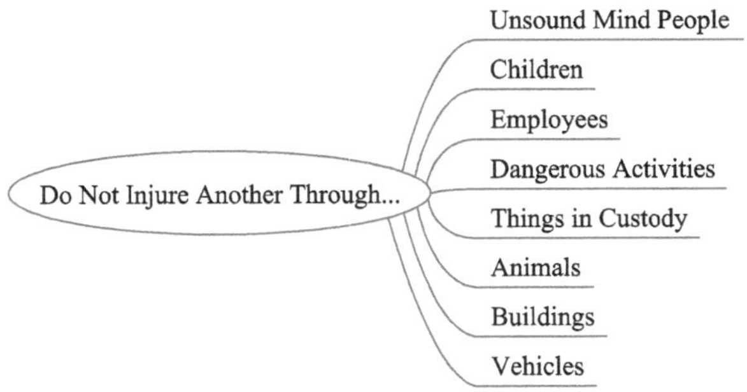

Fig. 5.2 A civil law approach to the law of Torts

Code consequently specifies the individual's responsibility according to the subject matter: liability for the behaviour of other agents, dangerous activities, etc. For the sake of brevity, this tort liability regime may be summarized as seen in Fig. 5.2:

Here, it suffices to pay attention to Articles 2048 and 2052 of the Italian Civil Code, that is, liability for harms caused by an individual's children or animals. In both cases, contrary to the U.S. regime of tort liability, strict liability is established for Italian parents for every injury or harm caused by their children and animals, *i.e.*, regardless of whether the parents were aware of their children's propensity to cause a particular type of harm, whether the animal was a dangerous beast or a domestic pet, and so on. Aside from hard cases involving matters of legal causation, what the plaintiff has to prove concerns a "legally sufficient condition" between the behaviour of the agent, for which the Italian parents bear responsibility pursuant to Articles 2048 and 2052 of the Civil Code, and the actual loss or damage suffered by the plaintiff: your fourteen-years old kid broke my wife's sixteenth century Delftware vase, your pet bit my child, and so forth, with all the possible legal variants of the *commedia dell'arte*.

However, the Italian Civil Code also provides at the same time limits for such no-fault liability by reversing the burden of proof. On one hand, parents evade responsibility when they demonstrate that they could not prevent their child's action. On the other hand, owners or keepers of animals have to show that a fortuitous intervening event occurred. Admittedly, the devil is in the details and it is not simply a question of bringing such evidence before courts. Dealing with responsibility for the behaviour of my child, I should prove, for example, that the (*Picciotto Roboto* of the) local Mafia kidnapped me and therefore, I could not prevent my toddler from accidentally burning down your house last night. Even more difficult is the proof of a fortuitous event: a lightning struck the chain of the dog in the garden of my villa, freeing it so that the animal could escape and bite my neighbour in the roundabouts. Still, when compared with the American model of strict liability claims for the use of AI children and pets, the Italian way of mitigating such strict liability rules has its merit. Let me suggest three motives.

First, in order to sever the chain of responsibility, we have to pay attention to the circumstances and events surrounding the defendant, rather than the unpredictable behaviour of the robot. As stressed by Chopra and White (2011: 135), "in a world where artificial agents are not accorded legal personality, the act of an artificial agent, whether or not considered a legal agent, cannot 'break the chain of causation' and cannot be a proximate cause of injury in its own right." By placing on owners and users of robots the burden to prove that a fortuitous event, or a set of circumstances, broke the chain of legal causation, we can thus prevent some drawbacks of the U.S. model of tort law. In fact, for many years ahead owners and users of robots will not be able to understand exactly when a specific machine shows a dangerous propensity or trait not characteristic of that robotic model, so that defendants will hardly be able to prove they could not reasonably foresee any risk of harm to third parties. *Vice versa*, by following the Italian model, there will be a growing number of cases where owners and users of robots can evade responsibility despite the lack of foreseeability of the harm or the unpredictability of the robotic behaviour. The point here concerns the irresistibility of the event, or the set of circumstances, that resulted in individuals failing to prevent robots from harming others. Such scenarios can be likened to the ways some legal systems establish responsibility for the set of dangerous activities: individuals are not liable when there is evidence that they have taken all the "appropriate measures" in order to prevent damage.

Secondly, by focusing on the events or circumstances that may break the chain of legal causation, attention should be drawn to further cases of apportioned liability. This hypothetical is closer to parents who evade liability for the behaviour of their children in Italy, than responsibility for damages caused by animals. The analogy with the Italian approach to cases of negligence-based liability suggests that the plaintiff has to prove that a legally sufficient condition exists between the action of the robot and the actual loss or damage suffered by the plaintiff. On this basis, according to Article 2048 of the Italian Civil Code, defendants should prove that they could not prevent the harmful behaviour of the robot because, say, the negligent or intentional behaviour of the plaintiff prevented them from doing so. The rationale for this tort liability policy has been stressed time and again. In *Agency Law and Contract Formation* (2004), Eric Rasmusen shows a number of cases where third parties, rather than individuals bearing responsibility for the care of other agents, are in the best position to prevent harm or damages, so that third parties should be reckoned as "the least-cost avoider." Similarly to the effects of errors and malfunctioning of robots in the field of contractual obligations as seen above in Sect. 4.3.2, we may envisage cases where a third party should have been aware of the erratic conduct of the robot due to its ostensibly defective, or faulty behaviour, as the robot of Asimov that appeared to be "drunk." In these cases, defendants could argue that the negligent or even intentional wrongdoing of the third party caused or at least, contributed to the harm induced by the machine.

Finally, by setting limits to strict liability rules through the reversal of the burden of proof, this approach to extra-contractual obligations prevents a new Hegelian night where all types of tortious responsibility for the behaviour of robots turn out to be grey: see above in Sect. 4.5.2. In order to discern these multiple types of harms, the Italian Civil Code provides for different ways through which individuals evade responsibility for actual losses or damages caused by their animals, children, vehicles, dangerous activities, and so forth. Analogically, in the case of robots, we should distinguish between robots-as-means of human industry and robots-as-agents in social life. In the case of robots-as-means, *e.g.*, the industrial ISO 8373 robot mentioned in the introduction to Chap. 4, it seems fair to apply traditional rules of extra-contractual obligations such as strict products and malfunctions liability. Yet, dealing with robots-as-agents, such as service machines for personal and domestic use, it is a tricky question as to how we should grasp the potential harms. All in all, should harm caused by a robot toy or a robot nanny be likened to the responsibility of Italian parents for harm caused by their children, so that individuals could evade liability when it is proved they could not prevent the harmful conduct of the robot? Conversely, should the legal system tighten the burden of proof, by conceiving robots as the Italian Civil Code governs the behaviour of animals, so that individuals could evade responsibility only when they show that a fortuitous event occurred? But, how about the idea of considering robots as an individual's workers and employees, such as i-Jeeves 2.0, *i.e.*, the service robot for personal business illustrated in Sect. 4.4.1?

Indeed, reflect on this threefold scenario:

(a) A robot toy that spends most of the time at home playing with your children, and which, now and then, goes with them to the public garden accompanied by your robot nanny;
(b) The robot nanny that, going back home with your children and the robot toy, after having accompanied them to the public garden, stops at the mall to buy some milk and candies;
(c) i-Jeeves 2.0 that manages and makes use of the property for your family business, so as to pay bills, entering into binding contracts, hiring robot nannies, buying robot toys, and so forth.

The diversity of such robotic applications calls for different kinds of liability for harms in tort law. From a legal viewpoint, whereas the metaphor of a robot toy as a smart artificial animal, or even an AI child, suggests new cases of negligence-based responsibility for the behaviour of others, we may guess whether liability for both i-Jeeves and robot nannies should be likened to traditional types of responsibility for the behaviour of workers and employees. Here, tort liability depends neither on intentional wrongdoing nor on lack of due care, but on vicarious responsibility that does not admit limits by reversing the burden of proof. In the light of this further analogy, *i.e.*, the parallel between robots and workers, the focus is on the strict liability of humans for harm caused by a new generation of AI employees.

5.3 AI Employees and Strict Liability Rules

We have examined two cases where liability is established notwithstanding any illicit or culpable behaviour: strict products and malfunctions liability concerning robots as instruments of human industry in Sect. 4.2.2, and strict liability for robots as agents of human interaction, reckoned either as dangerous animals in Sect. 2.2.2, or according to the responsibility of Italian parents for the behaviour of their children and pets in the previous section. Yet, most legal systems provide for a further kind of strict liability, which fits this part of the laws of robots as agents of human interaction, specifically, the employer's liability for any illicit action the employees engage in under their work contract activities. Figure 5.3 illustrates such different types of strict liability for the behaviour of robots:

Let us now restrict the focus of this analysis so as to deepen that which American common lawyers sum up as the doctrine of *respondeat superior* and civil lawyers examine with clauses of strict liability, such as Article 2049 of the Italian Civil Code. Contrary to the hypothesis of strict liability for the behaviour of children and animals, neither the Italian Civil Code nor the

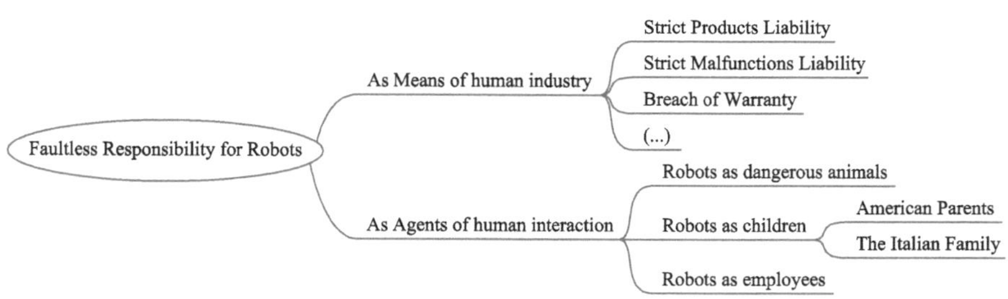

Fig. 5.3 Strict liability for robots in the law of Torts

American legal system provide limits to such no-fault liability. The reason for this hinges, on one hand, on the hierarchical subordination of the employees and the legal powers of the employers. For example, according to Articles 2104 and 2105 of the Italian Civil Code, employees have a duty of diligence, faithfulness and obedience. Conversely, among the powers of the employers, they have the right to direct, control and discipline their employees.

On the other hand, especially as articulated by American scholars, such a way of distributing social risk and responsibilities is justified on economic basis. In *A Legal Theory for Autonomous Artificial Agents* (2011: 128–129), quoting the sixth section of Posner's *Economic Analysis of Law*, Chopra and White affirm that "the economic rationale for strict liability rules like *respondeat superior* is best explained in terms of incentives on defendants to alter the rate at which they undertake particular kinds of activity. Courts applying a negligence standard typically examine how carefully a particular kind of activity is carried out, but do not question the level at which that activity is engaged in the first place. Strict liability addresses that need, for potential injurers subject to strict liability can be expected to take into account possible changes in activity levels and expenditures on care, in deciding whether to prevent accidents." In addition, dealing with damages caused by employees during work, the strict liability of employers guarantees third parties that such extra-contractual obligations would be met. Since most of the time employees lack the resources to cover the damages caused by their actions, they would not necessarily be responsive to the threat of tort liability. As Leon Wein argues in *The Responsibility of Intelligent Artefacts* (1992), the legitimacy of vicarious liability is "not grounded on a logical interconnection binding the wrongdoer to a loss he has brought about, but instead on a policy of providing compensation for loss, rather than imposing liability on financially incompetent parties. Consequentially, employers are answerable for their employee's autonomous acts even though they neither immediately influenced nor participated in the wrongful behaviour that occasioned the loss" (*op. cit.*, 110).

A tricky part of this framework concerns the link that must exist between the harm caused by the employee and the fact that the employee caused such harm under her work contract activities. In order to mitigate the regime of vicarious responsibility, for instance, Italian courts require that this link be understood as a matter of "necessary occasion." Back to the field of robotics, however, we can hardly imagine a service machine not undertaking its work activities. Lest lawyers embrace Sci-Fi scenarios, a court would never admit employers claiming that their robot caused harm, once finished its work duties and say, spending some free time with other robots at the coffee-shop. Besides, contrary to the duty or interest of contractual counterparties, third parties in tort law do not have to ascertain whether such a robot was actually behaving within its legal authority. Therefore, under strict liability rules for vicarious responsibility, owners and users of robots would be held strictly responsible for the behaviour of their machines 24-h a day, whereas, at times, negligence-based liability would add up to (but never avert) such strict liability regime.

This conclusion is harsh and once again, this could prevent individuals from buying and using robots in the first place. Strict liability rules of vicarious responsibility are even stricter than the strict liability rules for damages caused by dangerous animals or Italian children. In these latter cases, we have seen how no-fault responsibility is mitigated by reversing the burden of proof, so that owners and users of robots are not liable when the dangerous propensities of the robot were reasonably unknown, a fortuitous event occurred, humans could not prevent the harmful behaviour of the machine, etc. Yet, as Chopra and Write correctly remark in *A Legal Theory for Autonomous Artificial Agents* (2011: 130), "to apply the *respondeat superior* doctrine to a particular situation would require the artificial agent in question to be one that has been understood by virtue of its responsibilities and its interactions with third parties as acting as a legal agent for its principal." In other words, this strict liability regime would not fit all robotic applications but, rather, that special types of machines as examined in Sect. 4.5.1, *e.g.*, robots-as-agents in civil law. Let us explore how we should tackle this scenario of tort liability separately.

5.3.1 The Digital Peculium Revisited

Among the panoply of domestic and personal uses of robots for service tasks, such as entertainment, handicap assistance, personal transportation, or home security and surveillance, we examined the class of service robots for personal and professional businesses in Sect. 4.3. Although risky, such robots can be extremely useful in making contracts, or establishing rights and obligations between humans. In light of the business carried out by a new generation of robo-traders, as stressed the contracts made by such machines are valid. Moreover, it is feasible to strike a fair balance between the different human interests involved through new forms of legal accountability, such as the digital *peculium*. By employing robots to make business, transactions, or contracts, individuals could assert a liability limited to the value of their own robots' portfolio, while the *peculium* guarantees to the contractual counterparties of robots that obligations would really be met. In the field of torts, however, we are confronted with a much more complex scenario, in that rights and obligations established by robots do not simply concern their contractual counterparties and even any third parties involved by such contracts, *e.g.*, insurance companies. Rather, in the hypothesis of harms caused by robots as intermediaries of human life, the spectrum of third parties widens, so as to potentially include every human fellow, or other robot, that meets that robot by chance: in the case of unlawful or accidental damages caused to others because of the robotic behaviour, who is to pay?

The traditional viewpoint holds individuals strictly liable on the basis of, say, extra-contractual obligations for vicarious responsibility illustrated in the previous section. In order to mitigate such strict liability rules, owners and users of robo-traders could take out insurance much as traditional employers do. Aside from the technicalities of these insurance policies, the overall idea is that the insurance company does not only pay out when injuries occur in the workplace but, moreover, when the employer would be held responsible for injuries caused by her robotic employee. This scenario brings us back to the economic rationale for strict liability rules as incentives for employers to modify the rate at which they undertake their business via robotic agents. Whereas insurance premiums add to the costs of an individual's business through robots, the more such machines become reasonably safe and controllable, the more individuals will accept the risk of their use, notwithstanding clauses of vicarious responsibility.

However, we can improve this approach to the laws of robots in a twofold way. On one hand, we might extend the mechanism of *peculium* by determining that human strict liability should be limited to the value of their robot's portfolio or alternatively, guarantees of the *peculium* added to insurance contracts. Consider, for example, the model set up by the Rome Convention from 7 October 1952, on damages caused by foreign aircrafts to third parties on the surface. Although the application of this international Convention is based on the aircraft operator's strict liability, it provides for a limited compensation scheme for incidents as well as limits to such strict liability regime, by reversing the burden of proof. Similarly to the extra-contractual responsibility of Italian parents as examined above in Sect. 5.2.2, Article 6.1 of the Rome Convention establishes that:

> [A]ny person who would otherwise be liable under the provisions of this Convention shall not be liable for damage if he proves that the damage was caused solely through the negligence or other wrongful act or omission of the person who suffers the damage or of the latter's servants or agents. If the person

> liable proves that the damage was contributed to by the negligence or other wrongful act or omission of the person who suffers the damage, or of his servants or agents, the compensation shall be reduced to the extent to which such negligence or wrongful act or omission contributed to the damage.

In the event we decide to stick to a strict liability model of vicarious responsibility in the case of robo-traders, Article 11 of the Rome Convention suggests how we should interpret the idea that human strict liability can be limited to the value of a robot's *peculium*. In the case of the Rome Convention, the amount of financial compensation to be paid is determined on the basis of the weight of the aircraft causing the damage. In the case of robots, the amount of the *peculium* could be established on the basis of the "work contract activities" of the machine, so as to distinguish between, say, the duties of a robot nanny and those of i-Jeeves.

On the other hand, we can even stretch the original mechanism of *peculium* further so as to conceive robots, similarly to traditional artificial persons, as proper agents in business and civil law. As mentioned in Sect. 4.5.1, several scholars have endorsed this idea because the personal accountability of robots would simplify a number of contentious issues, such as whether robots are acting beyond certain legal powers, which party should be held liable for conferring such powers, or whether humans can evade liability for possible malfunctions of a machine. By recognizing the personal accountability of robots, we prevent, in other words, the intricacies of adding a new hypothesis of extra-contractual obligations for the behaviour of others, *i.e.*, animals, children and employees, in that (some types of) robots would be directly liable for provoking an injury and an actual loss or damage to third parties. In such cases, the *peculium* of the robot guarantees that extra-contractual obligations would be met, regardless of whether a human being should be held strictly liable, or deemed as negligent. All in all, this framework "provides a more complete analogue with the human case, where a third party who has been deceived by an agent about the agent's authority to enter a transaction can sue the agent for damages" (Chopra and White 2011: 162). Moreover, such "a more complete analogue" simplifies the complex mechanism of the burdens of proof examined throughout this Chap.. Although we may envisage further futuristic scenarios, such as robots as liable for the behaviour of other robots, *e.g.*, a robot nanny responsible for the conduct of a robot toy, as suggested above in Sect. 5.2.2, the legal mechanism of who has to prove what, namely, how the burden of proof works in the legal field, can properly tackle the challenges of technology. Whether dealing with artificial or natural agents the structure of the legal argumentation will likely remain the same.

5.4 Burdens of Proof

Problems of accountability and liability in the legal domain are intertwined with the mechanism of the burden of proof. According to the maxim of Roman law, *onus probandi incumbit ei qui dicit, non ei qui negat*, namely, the burden of proof does not fall on defendants, but rather on the party making allegations concerning a fact or legal issue. In criminal law, the burden falls on prosecutors to demonstrate that defendants are guilty on account of any action or omission prohibited by specific norms or statutes. In contracts, the burden falls on the parties alleging a breach of the agreement by their own counterparties. In tort law, the burden falls on the plaintiff who has to produce evidence of the defendant's wrongdoing as the cause of the plaintiff's harm. Whilst tortious claims are traditionally differentiated in intentional wrongdoing, negligence-based liability and no-fault responsibility, a further distinction is necessary in the case of robotic torts. Since some of these machines act, as much as animals and human beings, robots raise a new type of human responsibility for the behaviour of others. In light of different kinds of extra-contractual obligations for the design, production, supply, and use of these machines, matters of tort liability suggest that we should distinguish between torts concerning robots-as-means of human industry and robots-as-agents in social interaction.

In the case of tort liability for robots-as-means, *e.g.*, the ISO 8373 industrial robots introduced in chapter 4, claims of liability mostly arise out of strict product and malfunction liability of designers, manufacturers, and suppliers of such machines. Setting aside cases of intentional wrongdoing and criminal prosecution, tort liability may concern cases of strict liability or negligence-based responsibility. How the mechanism of the burden of proof can be applied to the field of robotic torts in such cases can be summarized like this: first, in most legal systems, the default rule is given by strict liability norms. This means that the claims of the plaintiff depend on a legally sufficient condition between the problems with the robotic application and the plaintiff's damages under the strict liability regime. According to the jargon of U.S. common lawyers, there must be "evidence from which a rational finder of fact could find in his favour" such a causal link, regardless of any illicit or culpable behaviour of the defendant as discussed already above in Sects. 3.5 and 4.2.2.

Secondly, how this mechanism allocates to one party or another the obligation of gathering and presenting further evidence, depends on the legal system with which we are dealing, much as crucial differences between common law and civil law traditions, adversary and non-adversary systems, burdens of production and of persuasion, down to juries and judges in charge of managing the case.[4] However, it does not follow the absolute liability of defendants from a strict liability regime: most of the time, defendants can indeed prove that they have taken all the appropriate measures in order to prevent any sort of damage and, moreover, that such causal link between the problem with the robotic applications and the plaintiff's damages does not exist. For instance, defendants may demonstrate that the product was not defective after all, or subordinately, this defect was not the proximate cause of the plaintiff's injuries, or such defect arose after the product was beyond the manufacturer's control. In the case of strict malfunctions (as opposed to strict products) liability, defendants may also demonstrate an abnormal use of the robotic application, much as the existence of reasonable secondary causes for the accident, etc.

Thirdly, strict liability rules do not prevent further kinds of responsibility for designers, manufacturers, and suppliers of robotic applications. For example, regarding claims of negligence-based liability, the plaintiff can prove a duty to conform to a certain standard of conduct, so that because of a breach of that duty, defendants caused an injury and an actual loss to the plaintiff. This kind of responsibility may concern forms of apportioned liability between suppliers and manufacturers of robots, or forms of vicarious responsibility for the negligent behaviour of the defendant's employees, such as designers of robotic applications. In any event, such forms of liability add up to the strict liability regime mentioned above.

On the other hand, there are cases of tort liability for robots-as-intermediates or proper agents in the civil law field. As occurs with the first class of robots, *i.e.*, robots-as-means of human industry, plaintiffs have the burden to show a legally sufficient condition between the problems with the robotic application and damages caused by such machines under the strict liability regime. In addition to strict malfunctions or products liability, however, this second class of robots raises a further set of cases of tort liability, where

---

[4]Contemplate for instance the difference between a non-adversary system, where judges might simply go on looking for evidence until the court satisfied itself that no such evidence exists and therefore that the factual claim or defense does not exist in that particular case, and adversary systems, where such unlimited searching is not permitted. Moreover, in US law, the burden of production has become decreasingly important as modern civil procedure expands "discovery," *i.e.*, the set of tools that allows one party to gain evidence by simply asking an opposing party for it. In addition, the burden of persuasion functions as a tiebreaker rule in US law, in such rare cases where the jury cannot make a decision because the evidence is equally divided. Contrary to most European legal systems, this power of the jury to decide facts, and even to decide if the facts are in equipoise, on the basis of persuasion-burden rules, partially negates all the "morality choosing" that legislators have considered in making their legal rules.

defendants are strictly responsible for the behaviour of others. Here, it is likely that the burden will mostly fall on the user, rather than the manufacturer or supplier, of such machines. Regardless of whether the case will concern negligence-based responsibility or strict liability, the mechanism of attributing to the parties the burden of proof varies with the analogy we endorse. The parallel between robots and employees, children or animals, sheds light on how responsibility for the behaviour of robots in tort law can be approached in the foreseeable future.

First, we may compare robots for service and domestic use to AI employees: the vicarious liability of the user would not let humans evade responsibility, once the plaintiff brings evidence of a legally sufficient condition. This is in agreement with the opinion of tort law scholars that consider either robots as dangerous animals, or their use as an ultra-hazardous activity. Strict liability rules thus apply to all the circumstances as seen above in Sects. 2.2.2 and 3.4.3.

Second, we can compare robots for personal or domestic use with children under the responsibility of their parent in American law, as illustrated in Sect. 5.2.1. Here, defendants need to prove their machine did not present any dangerous propensity or trait that is not typical of similar applications. Admittedly, for the foreseeable future, little room would be left for defendants to prevent liability.

Third, we may compare robots with children under the responsibility of parents as in Italian law. In this case, defendants avoid responsibility when evidence shows that they could not prevent the harmful behaviour of the robot, or that a fortuitous event occurred. Whereas some convergences with U.S. tort law model may emerge, the aim of defendants would still be particularly burdensome as highlighted above in Sect. 5.2.2.

However, legal systems could also endorse forms of limited liability such as the digital *peculium*. By applying this institution of Roman law to the field of extra-contractual obligations, strict liability for robots could be limited to the value of their portfolio or alternatively, guarantees of the *peculium* could be added to the clauses of insurance contracts. In addition, we can further stretch the analogy with the institution of Roman law of the *peculium* as a form of personal accountability for (some types of) robots. As aforementioned, several scholars support this idea, since granting "personal accountability" to robots would prevent the difficulties of a new hypothesis of extra-contractual obligations for the behaviour of others. Such "a more complete analogue with the human case" (Chopra and White 2011: 162) would not only hold robots directly liable for provoking an injury and any actual loss or damage to third parties. Moreover, the personal accountability of robots would fit a set of further cases concerning robots responsible for the behaviour of other robots, *e.g.*, cases of negligence-based liability or strict responsibility for the duty of a robot nanny to take care of the conduct of a robot toy as examined above in Sect. 5.2.2.

Yet, certain scholars find this scenario problematic, because even the best-intentioned and best-informed designer cannot foresee all the possible outcomes of robotic behaviour. Besides cost-benefit analysis and legal technicalities, such as the digital *peculium*, some stress the new "strong moral responsibilities" of designers and producers of robots due to the growing unpredictability of their behaviour (Grodzinsky et al. 2008). Others wonder whether the aim to produce robots with which individuals may bond is ethically justifiable (Dautenhahn 2007). Along with the employment of robots in battle and the increasing intricacy of network-centric applications, the use of "sensitive technologies" in the civil, as opposed to the military, field has suggested some scholars to invoke the "precautionary principle" (Veruggio 2006). Despite competing formulations, this principle basically states that we should reverse the burden of proof in order to prevent action when we are not (scientifically) certain that no dangerous effect would ensue. The legal terms of the principle can be further illustrated with the European Commission Communication from February 2000: "The precautionary principle applies where scientific evidence is insufficient, inconclusive or uncertain and preliminary scientific evaluation indicates that there are reasonable grounds for concern that the potentially dangerous effects on the environment, human, animal or plant health may be inconsistent with the high level of protection chosen by the EU."

Dealing with risks and threats that mostly depend on the unpredictable behaviour of robots and their impact on human health and the environment, we have to widen the focus of the analysis and consider how the enforcement of the precautionary principle may, partially or fully, affect an individual's rights and obligations. Four different ways in which the precautionary principle can legally be understood are examined next, along with how they affect the mechanism of the burden of proof. Then, the focus in Sect. 5.4.2 is on the limits of the precautionary principle in accordance with the alternative principle of openness. This analysis introduces the final chapter of this work on the law as meta-technology, and how legal systems may grasp the agenthood of robots.

5.4.1 The Precautionary Principle

The precautionary principle plays a key role in today's legal systems, addressing matters of harm, risk and scientific uncertainty that stem from the complexity of the issues with which we are dealing. In the wording of

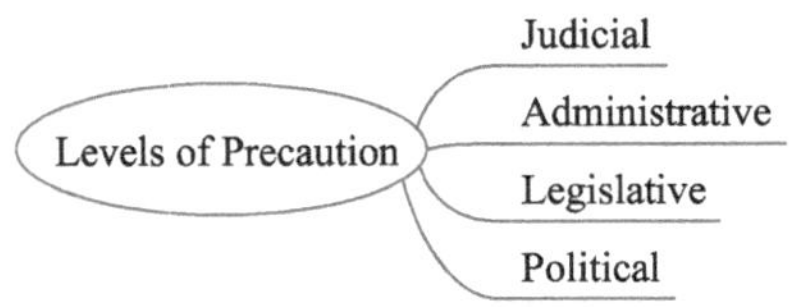

Fig. 5.4 Reversing the burden of proof with the precautionary principle

the International Commission for Electromagnetic Safety and the Benevento resolution from September 2006, every time "there are indications of possible adverse effects, though they remain uncertain, the risks from doing nothing may be far greater than the risks of taking action to control these exposures. The Precautionary Principle shifts the burden of proof from those suspecting a risk to those who discount it." More particularly, we should distinguish different levels of analysis that depend on how the reversal of the burden of proof is determined, namely, at the judicial, administrative, legislative and political levels of precaution as illustrated by Fig. 5.4:

By "judicial level" I refer to the adjudication by Courts. Under certain circumstances, tribunals can abandon the principle of *actori incumbit probatio*, so as to reverse the burden of proof and make it fall on the defendants of the case. This is what some parties have claimed in a number of lawsuits before the International Court of Justice. For instance, in *New Zealand v. France* decided in 1995 concerning the French nuclear tests in the Pacific Ocean, New Zealand's claim was that France should have proven the safety of its own activities. In the words of the petitioner in its *Request for an Examination of the Situation*, under the precautionary principle "the burden of proof fell on a State wishing to engage in potentially damaging environmental conduct to show in advance that its activities would not cause contamination" (§ 34). Likewise, in *Malaysia v. Singapore* from 2003, the advocate for Malaysia, Elihu Lauterpacht affirmed that "one may argue about the status of the precautionary principle, but Malaysia submits that this Tribunal should not reject the widely-held view that it is for the State that proposes action that may detrimentally affect the environment to show, not to itself, but to those that may be affected by it, that there is no real likelihood of harm to the environment" (quoted by Foster 2011: 247).

The second level of the precautionary principle concerns the regulatory powers of administrative authorities. Going back to the da Vinci surgery systems examined in Sect. 4.2, producers of such applications have to proactively demonstrate that the commercialization and use of robots for medical purposes is satisfactorily safe. On the basis of scientific evidence, Intuitive Surgical could thus obtain the authorization of the U.S. Food and Drug Administration, *e.g.*, approval Z-0658-2008 for "the Class 2 Recall da Vinci Surgical System 8 mm Long Instrument cannula." Likewise, in the

EU legal system, directive 93/42/EEC requires substantial clinical data guaranteeing the safety of medical devices. Producers of such devices, for instance, are to provide "a compilation of the relevant scientific literature currently available on the intended purpose of the device and the techniques employed" (Annex X, 1.1.1), the aim being to "determine any undesirable side-effects, under normal conditions of use, and assess whether they constitute risks when weighed against the intended performance of the device" (Annex X, 2.1). Similarly, in the case of UAVs, as seen above in Sect. 4.5, the burden of proof falls on producers and manufacturers of such unmanned aircrafts that should preventively demonstrate "their capability and means of discharging the responsibilities associated with their privileges." In the wording of Article 8(2) of the EU Regulation 216/2008 on common rules in the field of civil aviation and establishing a European Aviation Safety Agency ("EASA"), "these capabilities and means shall be recognized through the issuance of a certificate. The privileges granted to the operator and the scope of the operations shall be specified in the certificate."

A third level of the precautionary principle – "the legislative level" – concerns legal obligations established by national and international lawmakers. These obligations may involve the regime of legal presumptions, so that courts should reverse the burden of proof by applying such provisions, rather than using their own adjudicative powers, as seen in some cases of no-fault responsibility in this chapter. However, legislators can also establish such precautionary provisions on a "case-by-case" basis: under the World Trade Organization ("WTO") agreements, for example, Article 3.3 on the application of Sanitary and Phytosanitary ("SPS") measures entails a precautionary approach because it allows Members to adopt SPS measures which are more stringent than measures based on the relevant international standards, "if there is a scientific justification, or as a consequence of the level of sanitary or phytosanitary protection a Member determines to be appropriate in accordance with the relevant provisions of paragraphs 1 through 8 of Article 5." Similarly, under Article 12(1) of Regulation 258/97, EU law requires that Member States should have "detailed grounds" for considering that the use of a new kind of food endangers human health or the environment. As the Court of Justice of the European Union declared in *Monsanto v. Italy* (C-236/01) on 9 September 2003, "it follows that the reasons put forward by the Member State concerned, such as result from a risk assessment, cannot be of a general nature. Nonetheless, in the light of the limited nature of the initial safety analysis of novel foods under the simplified procedure… and of the essentially temporary nature of measures based on the safeguard clause, the Member State satisfies the burden of proof on it if it relies on evidence which indicates the existence of a specific risk which those novel foods could involve."

The final level of the precautionary principle regards the political choices that have to be taken. To date, the principle has concerned highly sensitive issues as the extinction of species, public health, food safety or global warming. The burden of proof should indeed fall on those who advocate taking action, because of the direct consequences on (a vital part of) the environment as a whole. By considering the threats and risks of robots, certain parties have consequently invoked the application of the principle to the field of robotics as well. For example, the *EURON Roboethics Roadmap* emphasizes that "problems of the delegation and accountability to and within technology are daily life problems of every one of us." Today, "crucial aspects of our security, health, life, saving, and so on" are conferred "to machines. Professionals are advised to apply, in performing sensitive technologies, the precautionary principle" (Veruggio 2006: 12).

The applicability of the precautionary principle to the field of robotics, however, raises three problems on how to deal with matters of uncertainty and ignorance. First, think of the threshold for applying the principle of precaution, namely, the existence and degree of scientific uncertainty as to the harm that the use of sensitive technology might invoke. In *Science and the Precautionary Principle in International Courts and Tribunals* (2011), Caroline Foster sums up a number of scholarly definitions concerning this level of risk: there should be a reason to believe that there are reasonable grounds for concern, a prudent belief in the cause or a plausible risk of harm, a credible threat or a non-negligible environmental risk, a likelihood of harm or reasonable possibility of damage. In the words of Foster, "the conclusion… is that there must clearly be some minimum threshold of scientific uncertainty in order for the precautionary principle to be applied. However, this threshold remains to be identified in practice" (*op. cit.*, 257).

Second, the precautionary principle may lead to irrational, protectionist, risk-averse or simply paradoxical outcomes. Consider the classic epistemological argument of Karl Popper's falsificationism, namely the assumption that, from a logical viewpoint, a scientific theory cannot conclusively be verifiable, although it shall conclusively be falsifiable (Popper 1935/2002). Hence, in the case of the precautionary principle, some have invoked a sort of "reversed Popperian paradox," since the need of proving the absence of risk before taking action, rather than proving the existence of such risk, implies that inactivity would continue until a no-evidence hypothesis is falsified. As Giovanni Rezza claims in *The Principle of Precaution-Based Prevention* (2006), "interventions for reducing potential risks of exposure to potentially hazardous sources should be implemented until the hypothesis is definitively proven to be false. Although the hypothesis would be in principle falsifiable, corroboration of the null hypothesis (*i.e.*, GMOs are unsafe), by definition, would never be satisfied, because of the early implementation of a ban. As in a reversed Popperian paradox, the intervention would continue unless/until the no-evidence hypothesis were falsified." According to this viewpoint, only independent research would be able to generate sufficient knowledge and empirical data in order to make rational decisions.

Third, lawyers properly speak in terms of burden of proof when they are dealing with some things that are, partially or entirely, uncertain. Nevertheless, we have seen throughout this work that the allocation of the burden of proof varies according to the field we are concerned with, and there are many cases where the precautionary principle is debatable. Indeed, there may be a strong rationale for engaging in action, so as to endorse what I would like to call here the "principle of openness": *Act despite of your own ignorance!* This is what really happened on 26 June 1997, when the U.S. Supreme Court struck down part of the *Communications Decency Act* ("CDA"), due to the particular nature of the means, *i.e.*, the internet. In the phrasing of Justice Stevens:

> In this Court, though not in the District Court, the Government asserts that – in addition to its interest in protecting children – its "equally significant" interest in fostering the growth of the Internet provides an independent basis for upholding the constitutionality of the CDA... The Government apparently assumes that the unregulated availability of "indecent" and "patently offensive" material on the Internet is driving countless citizens away from the medium because of the risk of exposing themselves or their children to harmful material.
>
> *We find this argument singularly unpersuasive.* The dramatic expansion of this new marketplace of ideas contradicts *the factual basis of this contention.* The record demonstrates that the growth of the Internet has been and continues to be phenomenal. As a matter of constitutional tradition, *in the absence of evidence to the contrary, we presume* that governmental regulation of the content of speech is more likely to interfere with the free exchange of ideas than to encourage it (*italics added*).

The precautionary principle should perhaps be applied to a number of applications in the field of robotics, *e.g.*, robot soldiers and squads of tiny drones that plan the mission they are to execute by themselves. Yet, in light of further applications such as NAO or the Japanese pop star robot singer HRP-4C, it seems clear that the precautionary principle does not offer a "one-size fits all" rule. Here, the burden of proof falls on those who want to prevent individuals from acting, so that scientists and companies should feel free to continue their research and business. In fact, precaution does not imply a ban on action because of ignorance, but an implementation to "act so that the effects of your action are compatible with the permanence of genuine human life" (Jonas 1979). In light of today's debate on whether we should employ such robots, as autonomous lethal machines and service applications for business purposes, domestic machines for edutainment, surveillance, and so forth, the aim is to explore in the final section of this chapter how the "imperative of responsibility" works in the laws of robots.

5.4.2 *Robotic Openness*

In between the extreme cases where robots imperil or, conversely, do not seem to affect Hans Jonas's "genuine human life," a significant grey zone of robotic applications illustrates how judgements can be particularly difficult. Consider if precaution should prevail for both robotic network-centric applications in the financial sector and semi-autonomous lethal weapons in battle. As stressed in Sect. 3.2, two different kinds of questions have to be distinguished concerning whether possible harm should impose a ban of a certain technology. On one hand, lawful uses of technology may depend on political decisions, as shown by the field of military robotics and current debate on whether lethal force can be fully automated, and what parameters, or conditions, should govern the use of robot soldiers: this is that discussed in Sects. 3.3.4 and 3.4.1. On the other hand, on the basis of scientific evidence and matters of legal causation, lawyers ascertain whether technology is capable of substantial lawful uses as examined above in Sects. 3.5, 4.2, 4.5 and 5.4.1. Here, the focus is on how the burden of proof should be allocated in these cases.

First of all, we have to pay attention to the political, rather than the administrative or normative level, of precaution. As seen in the introduction to Chap. 4 and in Sect. 4.1, the problem can be grasped in light of a spectrum. At one end, think of autonomous robot soldiers: in the name of precaution, it makes sense that the burden of proof shifts from those suspecting a risk in the design, construction and use of such machines, to those parties discounting the risk, so that political authorities should preventively demonstrate that their robots are reasonably safe and controllable. Whereas some robots have eventually been deployed without the necessary testing of the reliability of these autonomous weapons over the past years, the further employ of robots that cause serious harm by taking their own decisions could be interpreted as a war crime or a crime against humanity. *Vice versa*, at the other end of the spectrum, contemplate the case illustrated by other robotic applications, such as NAO or HRP-4C, that make it plain that openness, rather than the precautionary principle, should apply, since they do not impinge on what we may conceive of as a genuine life. However, in between such extremes, it is not a matter of equalizing pros and cons of openness and precaution. In addition to the philosophical reasons why we should endorse the ideals of the "open society" (*e.g.*, Popper 1945; Hayek 1960; etc.),[5] consider the legal reasons mentioned in the previous section, so that, notwithstanding risks and threats of robotic behaviour, research and development should continue, lest advocates of the precautionary principle share evidence that threats and risks outweigh potential benefits of this technology. After all, this is what advocates of the ban of (some classes of) robot soldiers aim to prove, so as to prohibit further work on a particular application, *e.g.*, squads of miniaturized autonomous lethal machines.

On this basis, attention should be drawn next to the limits imposed by the normative and administrative levels of precaution. Although, in accordance with the principle of openness, manufacturers of robots most of the time do not have to prove that the machines they are developing are risk-free, such robots still must comply with safety standards before their commercialization and use. This is that which was pinpointed in the previous section: in accordance with administrative authorizations and normative standards, such as the EU directive 93/42/EEC, the certificate will be issued only once it has been proven that the machine is safe. Whilst these standards vary according to the type of robotic application, *e.g.*, a robot-surgeon such as the da Vinci system, the reversal of the burden of proof means that, for every robot employed by humans, there must be evidence on the safety of the machine before it can be manufactured and employed by other human fellows. This leads to the level of the precautionary principle concerning the adjudicative powers of the courts and how matters of responsibility and the burden of proof should be determined.

Traditional cases of negligence-based responsibility in the field of torts were stressed in the introduction to this Chapter, by distinguishing between industrial robots and machines for personal or domestic use. Then, cases of negligence-based liability were scrutinized in Sect. 5.2 and with Figs. 5.1 and 5.2, by comparing the American with the Italian model. Next, cases of no-fault responsibility and more particularly, strict liability for the behaviour of AI employees were analysed in Sect. 5.3 and with Fig. 5.3. In light of such different types of tort liability, I have insisted in Sect. 5.4 on a key difference between responsibility for the behaviour of two classes of robots, that is, robots-as-means and robots-as-agents. Whereas current provisions on strict products liability, strict malfunctions liability, etc., properly address cases of liability for the use of industrial robots and generally speaking, robots-as-means of human interaction, a new generation of hard cases on tort liability for the behaviour of robots-as-agents is emerging.

[5] We return to these ideals below in Sect. 6.4.1.

No-fault liability for the behaviour of this class of robots can be mitigated with the right allocation of the burden of proof: yet, legal systems can also endorse forms of limited liability, such as the digital *peculium* and policies of compulsory insurance, in order to strike a fair balance in distributing responsibility and risk.

Admittedly, pros and cons of this latter standpoint were considered from an anthropocentric level of abstraction: the distinction between robots-as-means and robots-as-agents in fact leaves room for further questions, such as whether these machines should be conceived as autonomous legal persons with rights and duties of their own. Examination of this issue was postponed in Sects. 3.2, 4.3.3, and again in Sect. 5.2.2. Rather than focusing on the legal personhood of a robot toy, a robot nanny, or i-Jeeves 2.0, the attention has been on issues of human responsibility for the conduct of these machines. Accordingly, we have dwelt on robots as sources of human liability or conversely, as agents in civil law, rather than candidates for say, a new generation of constitutional rights and duties. Whilst a number of political choices will have to be taken, so as to choose between strict liability policies and negligence-based forms of liability, and how to balance precaution with openness, many of these decisions will concern the type of legal agenthood that a robot should have. After the analysis of criminal law, contracts and torts, the inquiry in the laws of robots has thus to be complemented with principles of constitutional law and the level of abstraction that is defined by the idea of the law as a meta-technology. The next chapter explores what type of legal agenthood robots should have through the set of notions and ways of legal reasoning with which the aim is to govern technological innovation.

Chapter 6
Law as Meta-technology

You'll love it! It looks just like a TeleFunken U-47

Frank Zappa, Joe's Garage

Abstract From the different classes of hard cases as mentioned in the previous chapters, it does not follow that the aim of the law to govern the process of technological innovation, necessarily falls short in coping with its own purpose. Yet, such hard cases on the legal personhood of robots, clauses of immunity, artificial agency in contracts, and new types of responsibility for the behaviour of others, raise the further issue on whether and how the existence and content of the law can always be determined on the basis of its own sources. Before the hard cases of today's laws of robots, the aim of this chapter is to determine which cases of robotics should be given priority and, moreover, whether one right answer is legally at hand, whether legal systems are open to alternative solutions, or political decisions need to be taken via international agreements. In light of the current debate on whether a certain type of drone design should be considered legal in the field of military robotics technology, for example, a reasonable compromise on the basis of legal expertise is at stake.Whereas both the UN General Assembly and its Secretary-General Ban Ki-Moon have been quiescent up to the date of publication of this book, it is noteworthy that the condition of immunity for the use of robot soldiers today goes hand in hand with no-fault responsibility for the employment of both industrial and service robots in the civil sector.

U. Pagallo, *The Laws of Robots: Crimes, Contracts, and Torts*, Law, Governance and Technology Series 10, DOI 10.1007/978-94-007-6564-1_6,

We can extend to the legal field that which Aristotle suggests about the notion of "being" in *Metaphysics* (VII 1, 1028 A 10): "There are many ways in which the law is said Aristotle (1984)." Throughout the centuries, the law has been conceived of as a form or a set of institutions, a structure or a superstructure, a function or a procedure, a tool for social control or an instrument of social communication. By considering the sources of the law, jurists further distinguish between political planning and spontaneous orders, statutes and customs. A short survey of comparative law reminds us of the differences between the civil and common law traditions, between the supremacy of the Code in continental Europe and the judge-made law of the Anglo-Saxon legal systems. In addition, different schools, such as the classical and the modern natural law tradition, legal realism and the Law and economics perspective, old and new kinds of institutionalism as well as several variants of legal positivism, such as inclusive and exclusive positivism, imperativism and normativism, aim to unveil the essence of the law. Although this variety of standpoints can be confusing and even disturbing, an analogy with the mathematical phenomenon of incompleteness may help to explain the current state-of-art. The law is said in many ways because the legal phenomenon is far more complex that its own language. According to Friedrich Hayek's remarks in the first volume of *Law, Legislation and Liberty* 1973, "I doubt whether anyone has yet succeeded in articulating all the rules which constitute 'fair play,' for example" (Hayek 1973: 76). When the aim is to define the essence of the law, answers require more information than that conveyed by the very question.[1]

The focus of this chapter on the law as a meta-technology does not suggest another version of what the law is, or of how it should be. The idea is to set the proper level of abstraction in order to understand the ways legal systems address the challenges of technological advancement and innovation. As mentioned in Sect. 2.1.3, a level of abstraction can be grasped as an interface, constituted by a set of features representing the observables of the analysis. By conceiving the law as a means that inter alia determines the conditions of legitimacy for the design, manufacture and use of technological artefacts, this level of abstraction renders an analysis of the system possible, with a resulting a model. Two such observables were examined in Sects. 5.2.1 and 5.2.2, *i.e.*, bans as well as the regulative frameworks for the commercialization and use of technology defining the legal responsibilities of the agents in the system. Bans can be established on the basis of empirical evidence or, conversely, simple ideological biases. Whilst today's debate on

[1]This thesis draws on Gregory Chaitin's work (2005), as discussed in Lolli (2008) and Calude (2008).

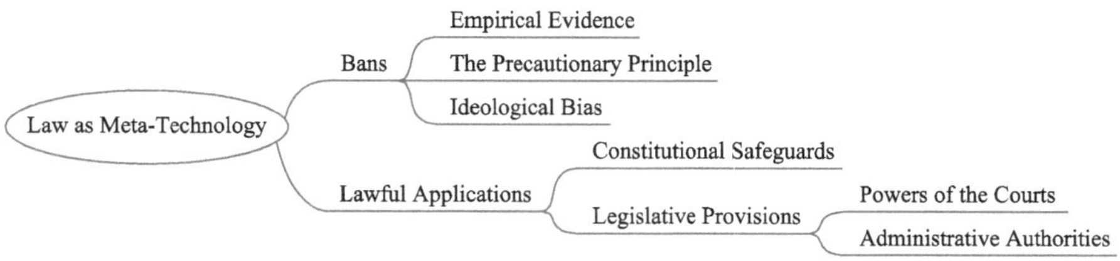

Fig. 6.1 Law and the challenges of technology

the precautionary principle illustrates how matters of empirical evidence and ideological biases may clash at times, we should pay attention to the ways society and its values impact on technology. From a legal viewpoint, once a certain technology is outlawed, the outcome is defined by the first step of the phenomenology of *Picciotto Roboto*: the simple use of this technology would be a crime, which adds to the prosecution of designers and manufacturers of the technology under the ban.

On the other hand, the regulative frameworks for the lawful commercialization and use of technology depend on both the constitutional safeguards of the system, if any, and the provisions adopted at national and international levels, such as 2001 Budapest Convention on Cybercrime. Further "variables" of the model were assessed in Sect. 5.2.2 in connection with the adjudicative powers of the courts, the supervisory powers of administrative authorities and how the burden of proof is allocated through the different steps of the legal process. Summing up the conditions where individuals are confronted with issues of legal responsibility, the complex network of concepts and ways of legal reasoning, with which the aim is to govern the advancement of technology, is illustrated with Fig. 6.1:

On this basis, four different types of cases have been analysed so far:

(a) Criminal uses of un/lawful robotic applications as examined in Sects. 3.4.2 and 3.4.3;
(b) Cases of immunity and affirmative defences in the criminal law field as pinpointed in Sect. 3.5;
(c) Cases of responsibility that depend on individual fault, *e.g.*, negligence-based liability, in both contractual and tort law as seen above in Sects. 4.2.2 and 5.2; and
(d) Cases of strict tort liability and how the burden of proof is reversed under such circumstances in both the law of contracts and torts, that considered above in Sects. 4.2.2 and 5.2.

Many scholars reckon, however, that the aim of the law, to govern the challenges of technological advancement and innovation, should be likened to the image of a turtle running after Achilles.[2] Think about the use of chlorofluorocarbons ("CFCs") since the 1930s and how legal systems needed half a century to outlaw the use of CFCs in our refrigerators. The opposite threat, of irrational risk-averse applications of the precautionary principle, was under scrutiny in Sect. 5.4.1: in a sort of reversed Popperian paradox, the need of proving the absence of risk before taking action can lead to a sterile inactivity. In between such extremes, the aim of the law to govern technology can nonetheless be effective, as shown by the *European Community (EC)-Asbestos* case under Article XX of the 1994 General Agreement on Tariffs and Trade ("GATT"). This article provides for environmental exceptions to the international covenant on free trade, establishing that the burden of proof falls on the party invoking such provisions, in order to demonstrate that restrictions are necessary to protect human health. After the French government passed a decree in December 1996 prohibiting the use of (products containing) asbestos and banning the import of such goods, Canada requested consultations with the European Community on 28 May 1998, so as to determine whether the French ban on chrysotile asbestos was compatible with Article XX (b) of GATT. On 18 September 2000, the World Trade Organization ("WTO")-Panel decided that the French decree did not fall within the scope of the technical barriers to trade ("TBT") of the international agreement. A few months later on 12 March 2001, however, the Appellate Body overturned the decision. Not only "prohibiting asbestos and asbestos-containing products had not been shown to be inconsistent with the European Communities' obligations under the WTO agreements," but the Appellate Body "reversed the Panel's finding that the TBT Agreement does not apply to the prohibitions in the measure concerning asbestos and asbestos-containing products and found that the TBT Agreement applies to the measure viewed as an integrated whole." Thanks to the regulative tools of the law, Europeans were no longer obliged to go on importing or employing asbestos.

The technicalities of the law, to be sure, do not prevent all the risks and threats brought on by the race of technology. The need humans have to adapt to the environment does not fade away with the development of today's complex systems and, moreover, such an evolutionary attempt can lead to the collapse emphasised by Jared Diamond in *How Societies Choose to Fail or Succeed* (2005). It suffices to mention the intricacies of the current debate on global warming. Still, contrary to traditional regulatory frameworks for the development and use of technological applications, a crucial peculiarity of robotic technology should be stressed. Besides robots-as-means of human

[2] See above in the introduction to Chap. 2.

industry, a further class of robots-as-agents was examined in Sects. 2.3.2, 4.4.1 and 5.4.2. Since such robots properly act as much in the same fashion as animals, children and adult human fellows,[3] it follows that robots should not only be reckoned as a source of responsibility in the legal field but, also, as agents of the system with some personhood of their own. In the phrasing of Chopra and White (2011: 189), "an artificial agent with the right sorts of capacities – most importantly, that of being an intentional system – would have a strong case for legal personality, a case made stronger by the richness of its relationships with us and by its behavioural patterns." As a result, focus should not only be on new types of responsibility that humans have for the behaviour of such machines but, also, on whether robots should be conceived as legal persons, or proper agents, of tomorrow's legal systems. This brings us back to Table 1.1 as seen above in the introduction to this book and Sect. 2.4.

The aim of this chapter is to fully analyse the legal observables of this model, by taking into account the conditions of responsibility for the behaviour of robots considered as legal persons, proper agents, or sources of damages in the legal system, that is, the three "Is," "SLs" and "UDs" of Table 1.1.

Next, I examine the debate on the legal personification of robots that has been particularly viral over the past years. Three normative positions are illustrated: the ability of individuals to have rights and duties of their own is distinguished from the ability to produce, through their intentional acts, rights and obligations that are binding on oneself or, conversely, on another. If acknowledging the legal personality of robots is not deemed a necessity, or even convenient, in the foreseeable future, three out of nine of the possible scenarios of legal responsibility would be excluded, namely I-1, SL-1 and UD-1 in Table 1.1.

The second section of this chapter draws attention to the ability of robots to produce, through their intentional acts, rights and obligations on behalf of humans. Although robots have no consciousness, free will or human-like intentions, the level of robotic autonomy is sufficient to have relevant effects in the civil (as opposed to the criminal) side of the law. Whilst an increasing number of scholars claim that robots should be welcomed as new agents in the field of contracts, cases of immunity, strict liability and robots' own responsibility for damages provoked by their fault, that is, cases I-2, SL-2 and UD-2 of Table 1.1 are considered in detail.

The focus in Sect. 6.3 is on cases I-3, SL-3 and UD-3. Rather than accountable AI agents out there doing business and entering into contracts, it is likely that legal systems most of the time will hold humans liable for the

---

[3] See above in Sect. 2.3.

behaviour of their machines. In addition to traditional forms of strict liability and negligence-based accountability, however, new types of responsibility can be envisaged. Think of new kinds of crimes committed by humans, who damage or destroy their robots in ways that are perceived as unjustified or disturbing by the community, as well as novel types of punishment for the behaviour of these machines. These latter injunctions are not directly addressed to, say, the owner of a robot; nevertheless, new punitive sanctions against the machine may affect its owner as well.

The final section of the chapter aims to prevent a possible misunderstanding. By conceiving the law as a meta-technology, it does not follow that technology does not impact on today's legal systems. No Sci-Fi is needed to admit that this is the first time ever the law will provide for the responsibility and agency of some artificial persons that are not reducible to an aggregation of human beings as the only relevant source of their action. By distinguishing robots as allegedly new legal persons, proper agents and new sources of liability, four out of nine possible cases of responsibility for the behaviour of robots should ultimately be judged under a legal strain: I-3, SL-2 and UD-2 and 3 of Table 1.1. In light of crucial differences between the fields under scrutiny, the aim is to pinpoint new scenarios of analysis and policy making. This introduces the final remarks of this book on the design of new environments for human-robot interaction.

6.1 Robots as Legal Persons

Scholars have increasingly been debating over the last decades whether legal systems should grant personhood to robots and, generally speaking, to autonomous artificial agents. This debate has involved legal experts as well as philosophers, sociologists, computer scientists and military experts. As Peter Singer reports in *A World of Killer Apps* (2011: 400), "today, the US Air Force has argued that its unmanned spy planes, if targeted by radar, have the same right to defend themselves with ammunition as its pilots have. This conferral on unmanned systems of the right to pre-emptive 'self'-defence makes sense from one perspective, but could also be a legal-dispute-turned-international-crisis in the making, as well as a huge (and probably unintentional) first step for the cause of robots' rights."

Advocates of the front of robotic liberation have obviously endorsed the idea that robots should have rights of their own. Moreover, this thesis has been partially supported by critics of the legal personification of robots. In *Rights of Non-Humans?* (2007), Günther Teubner insists, for example, on the risks that follow from the "socialization of things" and the fact that artificial agents act and decide beyond human control. This autonomy entails problems of alienation and reification of social relations that already troubled Karl Marx and Martin Heidegger. Still, according to Teubner, "multiple legal distinctions… have the potential to confer a carefully delimited legal status to political associations of ecological actants. And those real fictions may do their work as actors exclusively in institutionalized politics without necessarily appearing as actors in the economy, in science, medicine, religion or somewhere else in society. Legal capacity of action can be selectively attributed to different social contexts. The result is that law is opening itself for the entry of new juridical actors – animals and electronic agents" (*op. cit.*, 20).

By distinguishing between robots as proper agents in the legal arena, and robots as simple instruments of human interaction, the focus should be on the multiple ways legal systems may govern the new juridical actors. As tools of human industry, robots can be considered as subjects of clauses and conditions of contracts, sources of extra-contractual obligations, or innocent means in the hands of an individual's *mens rea*. *Vice versa*, by conceiving robots as agents in the legal field, far more complex scenarios should be taken into account. In light of Fig. 2.6 in Sect. 2.3.2, four different conditions of legal personhood were assessed. Theoretically speaking, legal systems might grant:

(a) Independent legal personhood to robots with rights and duties of their own;
(b) Some rights of constitutional personhood, such as those granted to minors and people with severe psychological illnesses, *i.e.*, personhood without full legal capacity;
(c) Dependent, rather than independent, personhood as it occurs with artificial legal persons such as corporations; and
(d) Stricter forms of personhood in the civil law field, such as the accountability of (some types of) robots for both contractual and extra-contractual obligations.

Naturally, we should examine the further legal variables of Fig. 2.6 so as to widen the perspective and take into account other forms of agenthood. Going back to Teubner's analysis in the *Rights of Non-Humans?*, the entry of new actors on the legal scene concerns all the nuances of legal agenthood, such as "distinctions between different graduations of legal subjectivity, between mere interests, partial rights and full-fledged rights, between limited and full capacity for action, between agency, representation and trust, between individual, group, corporate and other forms of collective responsibility" (*op. cit.*, 20). Let us dwell here on the canonical notion of "legal personhood" established by Article 1 of the 1948 Universal Declaration on Human Rights, so as to distinguish it from other forms of "restricted personhood" that are usual in the civil (as opposed to the criminal) law field. This perspective has been explored by Mireille Hildebrandt, Bert-Jaap Koops and David-Olivier Jacquet-Chiffelle in *Bridging the Accountability Gap* (2010), where they distinguish "legal persons who are capable of civil actions, such as contracting," from "legal persons who are capable of all types of legal actions, and who can bear both civil and criminal responsibilities; this is the category of legal persons who are also moral persons" (*op. cit.*, 550). Likewise, in *Cognitive Automata and The Law* (2009), Giovanni Sartor proposes a tripartite normative distinction that ends up in two kinds of personhood:

> To address the attribution of personality, we need to distinguish three normative positions:
>
> 1. the ability to have one's own legal position, *i.e.*, the ability to have rights and duties of one's own;
> 2. the ability to produce, through one's intentional actions, rights and obligations on one's head;
> 3. the ability to produce, through one's intentional actions, rights and obligations on the head of another.
>
> Only the first two positions characterise legal personality, broadly understood. The third one... is independent from the others: having legal personality does not entail that one is able to bind another; this usually presupposes a delegation by the concerned person (Sartor, *op. cit.*, 282).

As stressed above in Sects. 4.5.1 and 5.3.1, new forms of accountability for robots seem particularly fruitful in both contracts and tort law, since such approaches, *e.g.*, the digital *peculium*, simplify a number of controversial issues, such as robots acting beyond certain legal powers, matters of liability for conferring such powers, or whether humans should evade responsibility when the machine malfunctions. However, that which certain proponents are arguing is different. Forms of artificial accountability, such as the digital *peculium*, would not be unsatisfactory because the parallels between robots and, say, slaves are deemed unethical or anthropologically biased. Rather, the autonomy granted by such forms of accountability is reckoned insufficient because once we accept that some artificial agents may be properly conceived of as strict agents in the field of contracts, their legal personhood would follow as a result. In the wording of *A Legal Theory for Autonomous Artificial Agents*, "none of the philosophical objections to personhood for artificial agents – most but not all of them based on 'a missing something argument' – can be sustained, in the sense that artificial agents can be plausibly imagined that display that allegedly missing behaviour or attribute. If this is the case, then in principle artificial agents should be able to qualify for independent legal personality, since it is the closest legal analogue to the philosophical conception of a person" (Chopra and White 2011: 182).

This philosophical conception of a person is deepened next in Sect. 6.1.1: the aim is to ascertain whether today's legal systems should grant legal personality to robots. Then, the pragmatic, rather than conceptual, reasons why legal systems should acknowledge either the dependent or the independent versions of the legal personhood of robots will be discussed in Sect. 6.1.2. On this basis, the focus will be narrowed in Sect. 6.2 so as to ascertain whether we should welcome stricter forms of legal agenthood for these machines.

6.1.1 The Front of Robotic Liberation

Lawyers have discussed the meaning of "person" over the past two millennia. In the *Institutes* and the *Digest*, the word recurs in 168 different contexts of Gaius' work and comments. Although William Thorburn in *What is A Person?* (1917: 299) is probably right when asserting that "nowhere does he [Gaius] define or explain Persona," the word is often used in connection with a given individual (*e.g.*, *actio in personam*), the role of a party in a process or legal act (*e.g.*, *persona actoris*), the status of free men and slaves (*e.g.*, *persona sui iuris* and, conversely, *alieni iuris*), down to the distinction between the physical or "natural person" and the "personae vice fungitur" (*Dig.* 46.1.22). Likewise, Cicero, another famous Roman lawyer, employs the word to denote the party to a legal trial, as in *On the Laws* (*De Leg.* 2.48–49), or, according to the original meaning of the word, that is "mask." In addition, Cicero (1999) uses the word "persona" to define a character, a social role or function, the disposition or temperament of a person and, generally speaking, to stress the moral and spiritual features that mark an individual's "personality."

Admittedly, none of the Roman definitions of "persona" resembles, or anticipates, the current meaning of personhood as a legal subject with rights and duties of its own. For example, today's idea that a legal subject can be an "artificial person" should be traced back to the notion of "*persona ficta et rapraesentata*" developed by the experts of Canon Law since the thirteenth century. The classical definition of legal person that we find in chapter 16 of Thomas Hobbes' *Leviathan* has thus a precedent in the work of Bartolus de Saxoferrato (1313–1357). In his *Commentary on Digestum Novum* (48, 19; ed. 1996), Bartolus reckons that an artificial person is not really a person and, still, this fiction stands in the name of the truth, so that we, the jurists, establish it: "*universitas proprie non est persona; tamen hoc est fictum pro vero, sicut ponimus nos iuristae*." This idea triumphs with legal positivism and formalism in the mid-nineteenth century. In the *System of Modern Roman Law* (1840–1849) ed. (1979), Friedrich August von Savigny claims that only human fellows properly have rights and duties of their own, even though it is in the power of the law to grant such rights of personhood to anything, *e.g.*, business corporations, governments, ships in maritime law, and so forth.

On the other hand, Romans linked the notion of "persona" with that of human beings, including women and slaves. It is only with the Enlightenment, however, that the notion of "legal personhood" was intertwined with the ideas of equality and having rights, according to the "self-evident truth… that all men are created equal" (US Declaration from 1776), that "men are born and remain free and equal in rights" (Article 1 of the 1789 French Declaration), down to the 1948 Universal Declaration that "all human beings are born free and equal in dignity and rights."[4] Likewise, a legacy of the Enlightenment is the aim to rationalize the fabric of the law through the reform of criminal procedures and the systematization of codes. Think of the custom of placing animals on trial, which finally ended when human individuals remained the only plausible actors in the legal domain. This does not mean that the power of the law to grant rights to anything was formally overcome. Rather, the impulse to the rationalization of the legal system means that rights and duties of such artificial legal persons, such as corporations, governments or ships, should be reducible to an aggregation of human beings as the only relevant source of their action.

This ambivalence reverberates in today's debate on the legal personhood of robots, as shown by the seminal work of Lawrence Solum, *Legal Personhood for Artificial Intelligences* (1992). Here, Solum proposes "a thought experiment that may shed light on the debate over the possibility of artificial intelligence and on debates in legal theory about the borderlines of status or personhood" (*op. cit.*, 1256). The thought experiment there regards the Thirteenth Amendment to the US Constitution and whether it could legitimately be extended to (some smart) artificial agents. In order to determine whether legal systems should grant independent legal personhood to robots, Solum proceeds in a dialectical way, *i.e.*, taking into account three possible objections to the idea of recognizing rights to those artificial intelligences ("AIs"). As the Latin adagio says, *Veritas filia temporis*, the truth is the

[4] See above in Sect. 2.3.2.

daughter of time: as a son of his own era, all the objections considered by Solum have to do with the anthropocentric standpoint of today's legal systems. More particularly:

(a) "AIs Are Not Human" (*op. cit.*, 1258–1262). Drawing on the ideas of the Enlightenment, current legal systems have overcome prejudices and superstitions of the Middle Ages, finally leaving humans as the only plausible actor in the legal domain. Why should legal systems abandon their anthropocentric standpoint? What would the interest be in granting full legal personhood to robots? Whereas some scholars announced some years ago that intelligent machines will succeed humans and that we, as a species, would face extinction,[5] why on earth should we grant the rights of constitutional personhood to robots?
(b) "The Missing-Something Argument" (*op. cit.*, 1262–1276). Robots lack some critical elements of personhood such as consciousness, intentionality, desires and interests. According to the current state-of-art, robots thus lack the set of preconditions for attributing liability to someone in the field of criminal law. While criminal accountability and legal personhood are intertwined with the moral responsibility of the individual who has to be acknowledged as a legal person, a lawyer filing a civil rights action to ultimately convince the US Supreme Court that robots are entitled to the rights of constitutional personhood seems a hopeless case. Consider the responsibility of natural legal persons that depends on their reason and conscience, although humans may have rights without responsibilities due to severe psychological illnesses or emotional and intellectual immaturity.[6] On this basis, should we liken the personhood of robots to the rights of children or the insane?
(c) "AIs Ought to Be Property" (*op. cit.*, 1276–1279). Resting on the shoulders of John Locke's doctrine on property in §§ 25–51 of *Two Treatises of Government*, the argument is that robots are the product of human labour and, therefore, those who make robots are entitled to own them. Should Locke's thesis be the object of the same criticism he directed at John Filmer's paternalistic ideas in the *Patriarcha*? (ed. 1991). In other words, once robots can properly be reckoned as modern slaves, as seen above in Sect. 4.4, why should we emancipate them? Although, in the phrasing of Solum, "even slaves can have constitutional rights, be those

---

[5]See above in the introduction to Chap. 2, where the works of Moravec (1999) and Kurzweil (2005) illustrate this point.

[6]See above in Sect. 2.3.2.

rights ever so poor as compared to the rights of free persons" (*op. cit.*, 1279), what would such rights be? Would they concern "some measure of due process and dignity" (*ibid.*)?

Remarkably, there are no legal reasons or conceptual motives for denying the personhood of robots according to Solum: the law should be entitled to grant personality on the grounds of rational choices and empirical evidence, rather than superstition and privileges. Solum insists on this legacy of the Enlightenment, claiming that "interests and goods can be conceived as objective and public – as opposed to feelings, to which there is (at least arguably) privileged first-person access" (*op. cit.*, 1272). All in all, Solum's counter-arguments can be summed up with five points.

First, concerning the objection that "AIs are not human," we should preliminarily distinguish between dependent legal personhood, *e.g.*, corporations and the independent legal personhood of human fellows. New forms of accountability for the behaviour of robots, such as the *peculium*, are compatible with the anthropocentric standpoint of today's legal systems, as the level of autonomy insufficient to have robots found guilty by criminal courts, arguably is sufficient to have relevant effects in the field of contracts. A strict application of the Roman law mechanism of *peculium*, moreover, traces rights and duties of robots back to humans as the only relevant source of their actions, so that acknowledging the legal agenthood of non-humans, such as autonomous machines with *peculium*, menaces no pillar of today's legal framework.

Second, dealing with the "missing-something argument," all of the variants of this thesis depend on the notion of robotic intentions illustrated in the introduction to Chap. 3. Solum has a point when claiming that "if the practical thing to do with an AI one encountered in ordinary life was to treat it as an intentional system, then the contrary intuition generated by Searle's Chinese Room would not cut much legal ice" (*op. cit.*, 1269). Consider the field of civil (as opposed to criminal) law: John Searle may be right in that robots really do not understand what they are doing when they, say, randomly select from a uniform distribution of choices bids and offers under the constraint that they cannot intentionally lose money. From a legal viewpoint, however, what is crucial here is not the self-consciousness of the robot but, rather, whether such a machine can outperform humans, for example, in the double-auction experiments examined above in Sect. 4.3.1.

Third, the "missing-something argument" based on the uniqueness of humans is simply gratuitous as robots that can display such missing behaviour or attribute are highly imaginable. On one side, certain scholars claim that free will is a prerequisite of legal personhood and, yet, the thesis ends up in the conundrum of physical causation: "The most plausible story about human free will is that an action is free if it is caused in the right way – through conscious reasoning and deliberation. But in this sense, AIs also could possess free will" (Solum 1992: 1273). On the other side, contrary to the idea that legal personhood is subordinated to the capacities to experience emotions, desires, pleasures, or pains, Solum quotes one of the most distinguished advocates of the Enlightenment: "Kant's moral theory may cast some doubt on the assumption that emotion is required for personhood. Kant argued that all rational beings and non just humans are persons" (*op. cit.*, 1270). Although the philosopher from Königsberg could simply be wrong, Solum warns, "if human emotions obey natural laws, then (in theory) a computer program can simulate the operation of these laws... It should not be surprising that some AI researchers believe that an AI could (or even must) experience emotion" (*ibid.*).

Fourth, regarding the argument that "AIs ought to be property," Solum affirms that human nature is itself contingent and "we can imagine that in the distant future, scientists become capable of building the exact duplicate of a natural human person from scratch – synthesizing the DNA from raw materials. But surely, this artificial person would not be a natural slave" (*op. cit.*, 1278–9). It is not necessary to envisage a distant future, however, to show the weakness of the "property argument." After all, we already examined in Sects. 4.4.1 and 5.3.1 cases where people employ robots and, still, it is in their interest not to own them. The direct accountability of such machines can, in fact, strike a fair balance between the interest of robots' counterparties that both contractual and extra-contractual obligations would be met, and the claim of the users of such robots not to be dilapidated by the decisions of their machines.

Finally, the strongest argument against the legal personhood of robots is namely the thesis that "AIs *cannot* possess consciousness" (Solum 1992: 1264). This is indeed a crucial point since most scholars affirm that consciousness or, rather, self-consciousness, represents a key prerequisite of legal personhood. For example, in *Bridging the Accountability Gap* (2010), Mireille Hildebrandt et al. argue "that the relevant criterion is the emergence of self-consciousness, since this allows us to address an entity as a responsible agent, forcing it to reflect on its actions as its own actions, which constitutes the precondition of intentional action" (*op. cit.*, 558). Yet, if it is a matter of fact that today's artificial agents do not have this capability, Solum claims that nobody knows whether and to what extent tomorrow's robots will achieve it. In his phrasing, "I just do not know how to give an answer that relies only on a priori or conceptual arguments" (Solum 1992: 1264).

Advocates of the legal personhood of robots have not only asserted that none of the arguments against the full-fledged personality of such machines are consistent, but have even questioned the anthropocentric basis of today's legal framework. As Chopra and White affirm in *A Legal Theory for Autonomous Artificial Agents* (2011: 27), "the conditions for each kind of legal personality could, in principle, be met by artificial agents in the right circumstances. We suggest that objections to such a status for them are based on a combination of human chauvinism and a misunderstanding of the notion of a legal person."

Consider, for example, the "free will-argument." *Pace* Kant, recent findings in both neuroscience and cognitive psychology suggest that the idea of being sovereign of our own self, *i.e.*, Kant's notion of autonomy, is self-delusional. The objection that robots, contrary to humans, are "just a programmed machine" is rejected, because the combination of our biological design and social conditioning, on one side, and the programming of robots, on the other, suggest too many similarities "for us to take comfort in the proclamation we are not programmed while artificial agents unequivocally are" (Chopra and White 2011: 176). Along these lines, even the basic distinction between the moral accountability and responsibility of robots would fade away. Scholars should not only reflect on these machines as a possible source of relevant moral actions, that is, in the jargon of Luciano Floridi's "information ethics," the moral accountability of robots. Even though this latter perspective presents itself as an onto-centric, receiver-oriented, and ecological macroethics, so that the aim is to be "impartial and universal because it brings to ultimate completion the process of enlargement of the concept of what may count as a centre of moral claim" (Floridi 2008: 12), a step further would be necessary, to take into account the "moral sense" of robots seriously.

According to Chopra and White (2011: 166), "at the risk of offending humanist sensibilities, a plausible cause could be made that artificial agents are more likely to be law-abiding than humans because of their superior capacity to recognize and remember legal rules." Moreover, if such a law-abiding robot should break the rules, none of the reasons why legal systems currently punish people, such as deterrence, just deserts, education, or exemplary purposes, would be devoid of meaning. All the "perplexing questions" raised by Lawrence Solum (1992: 1247) could be properly met. As to the deterrence theory of punishment, Chopra and White claim that obedience to obligations can be embedded in the program of the machine, so that the robot would respond to the threat of punishment, by accordingly modifying its behaviour: in the words of *A Legal Theory*, "a realistic threat of punishment can be palpably weighed in the most mechanical of cost-benefit calculations" (*op. cit.*, 168). As to the "just deserts" function of punishment, the use of evolutionary algorithms and other mechanisms rewarding legal compliance or ethical behaviour would make the scenario of robots that understand why they should deserve some kind of reprimand realistic:

> The artificial agent's history of responding correctly when confronted with a choice between legal or ethical acts, whose commission is rewarded, and illegal or unethical acts, whose commission results in an appropriately devised penalty, would be *appropriate grounds for understanding it as possessing a moral susceptibility to punishment* (we assume the agent is able to report appropriate reasons for having made its choices). An agent rational enough to understand and obey its legal obligations would be rational enough to modify its behaviour so as to avoid punishment, at least where this punishment resulted in an outcome inimical to its ability to achieve its goals. While this may collapse the deterrence and just deserts functions of punishment, the two are related in any case, *for an entity capable of being deterred is capable of suffering retribution* (Chopra and White, *op. cit.*, 168–169, italics added).

In light of such cases of retribution and deterrence, it is not so hard to imagine the pattern of argument for the educative function of robotic punishment. Yet, even if I insisted in Sect. 5.2 on the legal relevance of how people in various ways train, treat, or manage their machines, for example, teaching a NAO robot how to play the violin, some differentiations should be maintained. In addition to the distinction between an individual's liability for the harmful behaviour of the robot, *e.g.*, a NAO damaging your 1721 "Lady Blunt" Stradivarius, and the robot's responsibility for its harmful behaviour, we should further distinguish between robots as targets of human censorship and robots that can be "forgiven" for their conduct (Chopra and White 2011: 180). There is indeed a difference between today's legal systems that order a dangerous animal to be eliminated and yesterday's trials against animals: the reason hinges on the need for differentiating the source of relevant moral actions, *e.g.*, a dog or robot killing someone, from the evaluation of such agents as being morally responsible for their behaviour. In the opinion of Chopra and White, however, "such rejections of personality for artificial agents implicitly build on the chauvinism – grounded in a dominant first-person perspective or in (quasi-) religious grounds – common to arguments against the possibility of artificial intelligence" (*op. cit.*, 172).

Returning to the Sci-Fi scenarios examined in Sect. 3.1, we should thus yield before the fact that, sooner or later, robots will be a sort of being *sui juris*, capable of sensitivity to legal obligations and even of susceptibility to punishment, insofar as such agents would be bestowed with the human-like equipment of free will, autonomy and moral sense. Still, it is debatable whether disagreement with the thesis of Chopra and White necessarily entails chauvinism or an obstinate form of anthropocentrism. After all, the starting point of the analysis should not be overlooked: Solum's thoughts concerning whether we should give robots constitutional rights raises a pragmatic, rather than logical, issue. Interestingly, this point is accepted by Chopra and White (2011: 154), in that "considering artificial agents as legal persons is, by and large, a matter of decision rather than discovery, for the best argument for denying or granting artificial agents legal personality will be pragmatic rather than conceptual."

From this common viewpoint, a restricted form of personhood for robots in the civil law field, such as the digital *peculium*, makes sense. This is a pragmatic way to strike a balance between the interests of robots' counterparties in that both contractual and extra-contractual obligations be met, and the claim of users or owners of such robots not to be financially ruined by the decisions of their machines.

In addition, since time is a scarce resource, a pragmatic viewpoint casts further light on what cases should be given priority in such fields as, say, criminal law. As stressed above in Sect. 3.2, a novel generation of offences, such as robot slavery and sex crimes against poor robot dolls, can be envisioned, so as to preserve consistency between robots and humans. Still, it is not a form of obstinate anthropocentrism or chauvinism to affirm that, nowadays, it is more urgent we address new cases of responsibility for the criminal behaviour of robots, than new forms of criminal accountability for humans that abuse their machines. Consider what the Satellite Sentinel Project reported on 10 April 2012 as to a British documentary film presenting evidence that the Sudanese government had committed crimes against humanity by bombarding civilians with some drones in the Nuba mountains of South Kordofan: "Surprisingly, the most irrefutable visual evidence comes from the Sudan Armed Forces, or SAF, in the form of video captured by a drone flown by SAF over apparent civilian areas in advance of bombardment. The evidence convincingly shows that the Government of Sudan is operating Iranian drones."[7]

However, some scholars argue that granting independent legal personhood to robots would provide for a more coherent picture of today's legal framework and, moreover, the legal personhood of robots and strict agency in contract law might be correlated. Accordingly, we should endorse what advocates of the front of robotic liberation claim, namely the independent, rather than dependent, legal personhood of robots, because this perspective simplifies several contentious issues in legal theory and "provides a more complete analogue with the human case" (Chopra and White 2011: 162). Therefore, let us deepen such theses in the next section: the aim is to weigh up the pragmatic reasons for conceiving robots as legal persons.

---

[7]Jonathan Hutson, *Sudan Armed Forces Implicated in Video Captured by their Own Drone*, retrieved at http://satsentinel.org/blog/sudan-armed-forces-implicated-video-captured-their-own-drone on 25 April 2012.

6.1.2 *The Pragmatic Stance*

There are two reasons why advocates of the legal personhood of robots claim we should agree with their stance, so as to carefully examine hypotheses I-1, SL-1 and UD-1 of Table 1.1. First, some affirm that strict legal agency in contract law and the legal personhood of robots might be correlated in order to prevent the ethical aberration of robots being treated as mere slaves. In *From Galatea 2.2 to Watson – And Back?* (2011), Mireille Hildebrandt suggests the following, "that for a computer agent to qualify as a legal agent it would need legal personhood. Both meanings of 'agency' raise questions as to the desirability of legal personhood of bots" and other AAs such as robots. Yet, it is not necessary to resort to the example of the legal status of slaves under ancient Roman law to show that forms of dependent or restricted legal status, such as agents in contract law, are not necessarily intertwined with forms of independent legal personhood. For example, the European Union existed for almost two decades without enjoying its own legal personhood. Moreover, in the case of robots, we need not grant them personhood as a way to prevent "the debates over slavery" that "remind us of uncomfortable parallels with the past" and "reflect ongoing tension over humanity's role in an increasingly technologized world" (Chopra and White 2011: 186). In fact, legal systems can determine new crimes committed by humans who unjustly damage or destroy their robots, regardless of the legal personhood of these machines. Whereas one solution could be to let the law charge humans for abuses of robots similar to those legal systems established for cases of animal cruelty in past decades, this does not mean that robots are capable of suffering, or they could experience emotions (Solum 1992: 1270). Rather, what is at stake here concerns the concept of that which may count as a centre of moral claims: as "informational objects," robots should indeed be considered as moral patients that deserve respect and protection as such (Floridi 2013).

The second argument of the front of robotic liberation is that granting personhood to robots would provide for a more coherent picture of today's legal framework. Admittedly, the parallels of robots with artificial persons would simplify a number of contentious issues in both the fields of contracts[8] and torts.[9] In *A Legal Theory for Autonomous Artificial Agents* (2011: 162), Chopra and White assert that "not only is according artificial agents with legal personality a possible solution to the contracting problem, it is conceptually preferable to the other agency law approach to legal agency

[8] See above in Sect. 4.5.1.

[9] See above in Sect. 5.3.1.

without legal personality, because it provides a more complete analogue with the human case." However, had not these same authors insisted on the thesis that the dependent, rather than independent, legal personhood of robots is "based on a combination of human chauvinism and a misunderstanding of the notion of legal person"? (*op. cit.*, 27) Why should we endorse an "analogy with the human case" in the case of robots that are criminally not accountable for their conduct? How could we improve the functioning of today's legal systems, by granting constitutional rights to robots?

This latter question brings us back to the thought experiment examined in the previous section with the thesis that "one cannot, on conceptual grounds, rule out in advance the possibility that AIs should be given the rights of constitutional personhood" (Solum 1992: 1260). Once a novel generation of robots endowed with human-like free will, autonomy or moral sense materializes, it would be reasonable that lawyers be ready to tackle a new generation of crimes, torts and contracts, including the proclamation of the constitutional personhood of robots. Still, there are two problems. On one hand, it seems reasonable to foresee that we should distinguish among the panoply of robotic applications. Whereas NAO the violinist and the Japanese pop star robot singer HRP-4C could be good candidates for constitutional personhood, it is hard to see the point of granting legal personhood to, say, an ISO 8373 industrial robot such as a manipulating machine for the manufacture of medical precision. Moreover, should we follow Peter Singer's suggestion that the unmanned spy planes of the US Air Force represent "a first step for the cause of robots' rights"?[10] In other words, should legal systems grant legal personhood to such autonomous and even intelligent artificial agents as the US Army's Global Hawk? I assume that advocates of robots as *sui juris* persons would admit to the nonsense in this conclusion.

On the other hand, if we do admit there being artificial agents capable of autonomous decisions "similar in all relevant aspects to the ones humans make" (Chopra and White 2011: 177), most scholars would acknowledge that, besides notions of crimes, contracts or torts, the meaning of person and that of legal personhood would change as well. In *Legal Personhood for Artificial Intelligences* (1992: 1260), Solum argues that, "given this change in form of life, our concept of a person may change in a way that creates a cleavage between human and person." Likewise, in *Bridging the Accountability Gap* (2010: 558–559), Hildebrandt et al. affirm that "it makes no sense to exclude outright non-human entities from such rights and responsibilities. His point [*i.e.*, Solum's] that such attribution should depend on the empirical

[10]See above in Sect. 6.1.

Table 6.1 Robots' behaviour and the "Factual Limits" of legal science

Responsible robot	Immunity	Strict liability	Unjust damages
As legal person	**Sci-Fi**	**Sci-Fi**	**Sci-Fi**
As proper agent	I-2	SL-2	UD-2
As source of damage	I-3	SL-3	UD-3

finding that novel types of entities develop some kind of self-consciousness and become capable of intentional actions seems reasonable, as long as we keep in mind that the emergence of such entities will probably require us to rethink notions of consciousness, self-consciousness and moral agency." However, nobody knows to where this scenario will lead. For instance, would an AI lawyer be an advocate of the tradition of natural law, a sort of legal realist or, contrary to the Kelsenian lesson of the pure doctrine of the law, focused on the substantive mechanisms of a new robotic order?

As a matter of fact, lest we revert to the imagination of science fiction writers, what the meaning of such legal concepts actually would be escapes the pragmatic grip of lawyers. As Wilhelm Leibniz used to say, "every mind has a horizon in respect to its present intellectual capacity but not in respect to its future intellectual capacity" (quoted by Allison P. Coudert 1995: 115). By drawing a line between the power of science fiction and the factual limits of legal analysis, we have to trace the boundaries of today's laws of robots in connection with the pragmatic issues of liability and responsibility for the behaviour of these machines. In the wording of *The Constitution of Liberty* (Hayek 1960: 23), "though we cannot see in the dark, we must be able to trace the limits of the dark area," concerning what we do not, or cannot, know. For the foreseeable future, it is thus likely that the independent personhood of robots will not be on the legal agenda. Although Sci-Fi approaches to the laws of robots often represent a fruitful way to address some legal challenges of this technology, as seen above in Sects. 2.1.1, 3.1, and 5.2.2, it is more than likely that the dependent, rather than independent personhood of robots, much as novel forms of responsibility for the behaviour of others in tort law, will have priority for pragmatic reasons. This conclusion can be illustrated with Table 6.1: leaving aside cases of immunity, strict liability and unjust damages for robots conceived as proper agents, or as sources of damage, which are examined in Sects. 6.2 and 6.3 below, Table 6.1 updates in bold the first row of Table 1.1:

Accordingly, three out of nine possible scenarios of legal responsibility illustrated in Table 1.1, namely I-1, SL-1 and UD-1, can be excluded on well-motivated grounds. By widening the spectrum of the analysis and considering the differences existing between the fields of crimes, contracts and torts, 9 out of 27 possible scenarios should thus be dismissed. These increase to 10 out of 27 if we consider the hypothetical of immunity and affirmative defences for robots as proper agents in the criminal law field, *i.e.*, another Sci-Fi scenario of the independent legal personhood of robots. On this basis, the time is ripe to pay attention to cases I-2, SL-2 and UD-2 of the model.

6.2 Robots as Strict Agents

Although in the foreseeable future robots will hardly be recognized as independent legal persons, *i.e.*, with rights and duties of their own, a number of reasons suggest taking seriously into account the "strict agency" of robots. As a matter of legal fact, agency and personhood are not equivalent, as the example of slaves in ancient Roman law and the status of the European Union from 1993 to 2009 confirm. From a pragmatic viewpoint, this makes sense, in that jurists should ascertain whether artificial agents can fulfil their duties and exercise discretion, rather if they can be aware of their own actions. In *Legal Personhood for Artificial Intelligences*, Solum dwells on this point through the "responsibility objection" and the "judgement objection" (*op. cit.*, 1244–1253). In his words, "we already have seen that making an AI a legal person [that is, an agent], a limited-purpose trustee, could have practical advantages, such as lower costs and less chance of self-dealing. The objection that the AI is not the real trustee seems to rest on the possibility that a human backup will be needed. But it is also possible that an AI administering many thousands of trusts would need to turn over discretionary decisions to a natural person in only a few cases – perhaps none" (*op. cit.*, 1254).

More than two decades after Solum's remarks, the personal accountability of robots in the field of contracts is supported by several scholars as a way of striking a balance between the interest of robots' counterparties to safely transact or interact with such machines, and the claim of users and owners of robots not to be dilapidated by the growing autonomy and even unpredictability of their behaviour. As stressed above in Sect. 4.4.1, new forms of accountability, such as the digital *peculium*, seem fruitful, in that such accountability renders irrelevant whether a robot is acting within certain legal powers, or who should be held liable for conferring such powers, whilst humans could evade responsibility for possible malfunctions of the machine, or errors of induction and specification. In addition to traditional mechanisms of distributing risk through insurance models and authentication systems, such forms of accountability might avert legislation that hampers the adoption of some useful applications, as the new generation of robo-traders, i-Jeeves and AI chauffeurs, as illustrated in Chap. 4.

In this context, let us reassess these ideas in connection with cases I-2, SL-2 and UD-2 in Table 1.1. Theoretically speaking, the focus should be on nine different scenarios, that is, immunity (I-2), no-fault liability (SL-2), and unjust damages (UD-2), involving robots as strict agents in criminal law, contracts and torts. However, only six of these cases, *i.e.*, I-2, SL-2 and UD-2 for both contractual and extra-contractual obligations are legally relevant, so long as robots are neither independent legal persons nor criminally accountable. Contrary to the traditional regulation of robots as sources of damages and responsibility for other agents in the legal system, that is, cases I-3, SL-3 and UD-3, what is at stake here concerns how the law may govern cases of liability through the forms of the direct accountability of robots. Lawmakers can obviously decide to establish the same type of rules for both cases, *e.g.*, robo-traders as well as robo-toys treated as simple sources of potential damages according to hypotheses I-3, SL-3 and UD-3. Yet, it would be meaningless to proceed the other way around, that is, robo-toys treated as robo-traders making contracts and business. What then is the specificity of cases I-2 (immunity), SL-2 (strict liability), and UD-2 (unjust damages), referred to robots as strict agents in contracts and torts?

To start with the hypothesis of immunity, this condition of irresponsibility can be illustrated with cases where robots should not be held to that which is impossible in the civil (as opposed to the criminal) law field. The tenet of the voidability of contracts between humans could apply to robot-traders pursuant to Article 1256 of the Italian civil code, Article 119 of the Swiss civil code, and so forth. However, aside from such borderline hypothesis, we should not miss a crucial point: while in the name of the principle of legality, a presumption of innocence is the default rule in criminal law, so that prosecutors have to prove that defendants are guilty on the basis of specific norms or statutes, immunity is the exception in civil law. It is thus difficult to imagine what kind of robotic activities should evade responsibility *a priori* by invoking the safe harbours of the law as strict agents in the field of contracts, much as the status of immunity concerning the internet service providers mentioned in Sect. 2.2.1. Indeed, we have to revert to the imagination of science fiction writers to envisage such cases where robots evade responsibility *a priori* doing business out there. This leaves four possible scenarios to be examined, that is, the hypotheses of strict liability (SL-2) and unjust damages (UD-2) for both contractual and extra-contractual obligations of these machines.

First, in the opinion of certain scholars, such as Curtis Karnow in *Liability for Distributed Artificial Intelligence*, a regime of strict liability can be imposed on the basis of a "Turing Registry" in the field of contracts (*op. cit.*, 193–196). Robots and other artificial agents, in other words, would be strictly responsible for harm or damages provoked by them as a way of coping with the growing unpredictability of their behaviour and, hence, the difficulty "to select out on a case-by-case basis the 'responsible' causes" (*op. cit.*, 191). By enlisting certified artificial intelligences, the Registry would insure owners and users of such agents against the risk of harmful behaviour, therefore striking a balance between the interest of both owners or users of robots to be protected from the unforeseeable conduct of their machines, and the claim of human counterparties to safely interact or transact with them. The higher the intelligence of the robot, the higher the risk, and thus, according to Karnow, the higher the risk, the higher the premium of the insurance (*ibid*).

Yet, it is not necessary to endorse such a strict liability policy as a "one size fits all" solution in the field of contracts. As stressed in Sect. 4.3.2, the intention of the robot is relevant when the legal effects of its contractual behaviour are under scrutiny, because humans do delegate to such machines cognitive tasks. Responsibility that depends on the fault of the agent allocates risk and liability for the behaviour of these machines in a more efficient way than rules of no-fault liability, since the legal effects of the cognitive states of the artificial agent should be assessed in light of the existing conventions of business and civil law. When the human counterpart had to have been aware of a mistake that, due to the erratic conduct of the robot, concerned, say, the substance of the agreement, it seems reasonable that humans shall not be able to avoid the usual consequence of such circumstances, that is the annulment of the contract. Conversely, a robot's counterpart should be allowed to expect, in good faith, that the machine really meant what it declared, *e.g.*, a contractual offer, so that the robot could not evade responsibility, claiming that it did not intend to conclude the agreement.

Admittedly, this form of responsibility stemming from personal fault looks more problematic in the field of torts. Although attributing accountability to robots can prevent several difficulties related to extra-contractual obligations for the behaviour of others, for the foreseeable future a strict liability regime would be more efficient. However, there are many cases where third parties, rather than individuals bearing responsibility for the care of other agents, are in the best position to prevent harm or damages, so that such third parties should be reckoned as "the least-cost avoider." As mentioned in Sects. 5.2.2 and 5.4, think about the third party that should have been aware of the erratic conduct of the robot due to its evidently faulty behaviour. For instance, defendants could argue that the

Table 6.2 A threshold of robots' responsibility in the civil law field

Robot behaviour	Responsibility	Immunity	Strict liability	Unjust damages
As legal person	In all the fields	Sci-Fi	Sci-Fi	Sci-Fi
As proper agent	**Contracts, Torts**	**Borderline**	**Why not?**	**Why not?**
Source of harm	(…)	(…)	(…)	(…)

negligent and even intentional wrongdoing of the third party provoked or, at least, concurred to the harm induced by the robot. Drawing on these arguments, our model can be updated. Leaving aside cases of immunity, strict liability and unjust damages for robots considered as sources of damage, which are examined in Sect. 6.3 below, Table 6.2 complements the Sci-Fi scenarios of Table 6.1 with the challenges to the laws of robots brought on by machines conceived as proper agents in the fields of contracts and torts. The conclusion is summed up in bold with the second row of Table 6.2:

So far, we have considered 18 out of the 27 possible scenarios of legal responsibility for the behaviour of robots. Ten of such hypotheticals were excluded in the previous section because of the criminal unaccountability of robots and their lack of independent legal personhood. By dismissing most of the hypotheses of immunity for contracts and torts, *i.e.*, I-2, four cases have been in focus in this section, that is, the direct liability of robots for contractual and extra-contractual obligations that Table 6.2 sums up in connection with cases of strict liability and unjust damages.

However, this framework is incomplete, since the personal accountability and responsibility of robots as strict agents in the civil law field does not exclude that such robots may have rights. Arguably, the growing autonomy and unpredictability of these machines prioritize issues of reliability and trustworthiness concerning their behaviour and still, this is not to say that the need of insurance mechanisms and further forms of guarantee should not be applied the other way around. Although referred to the independent legal personhood of artificial agents, Chopra and White (2011: 188) properly stress that "even in e-commerce settings, an important part of forming deeper commercial relationships will be whether trust will arise between human and artificial agents." Some, as Helen Nissenbaum in *Securing Trust Online* (2001), claim that trust would necessarily depend on shared norms and ethical values regulating social, that is human, interaction. Others affirm that trust does not necessarily entail an identifiable and direct human interaction and, moreover, as Cristiano Castelfranchi and Rino Falcone claim in *Principles of Trust for MAS* (1998), trust is feasible among artificial agents. By involving a decision to delegate and, furthermore, an expectation of gain by trust, this is nonetheless an issue that necessarily requires time, since positive outcomes grow so long as the act of trusting encourages more trust. This recursive effect explains why some intermediate solutions have been proposed over the past years, such as a “special normative system,” where robots hold rights and duties and enjoy full legal subjectivity. This status would not be directly recognized by the legal system and still, it would be binding for the parties to a contract. In the field of software agents, this is the mechanism Giovanni Sartor suggests in *Cognitive Automata and the Law* (2009: 283).

Once the wheels of this mechanism are oiled and the recursive function-effect of trust triggered, it is likely that such schemes will be progressively stretched, much as Roman lawyers did through the *peculium* of their slaves. A scenario where fully autonomous robo-traders employ assets and portfolios of their own fits the interest of owners or users of robots that the obligations of their robots’ counterparties would be met. Likewise, legal systems could set up forms of guarantees for the sake of robots and their own interests, *e.g.*, insurance policies that would pay out when a human establishes a tort against the robot, or that covers losses sustained directly by the machine. The example of the AI chauffeurs, illustrated above in Sect. 4.5.1, sums up this complex interaction of technological advancement, economic interests, political lobbying and legal mechanisms. Whilst the Governor of Nevada signed into law a bill authorizing the use of autonomous vehicles on public roads in June 2011, it is only a matter of time that the *peculium* of the AI chauffeur may be added to traditional forms of insurance. This portfolio would guarantee third parties against possible mishaps on the roadways, while insurance policies could similarly guarantee the robot in the case of accidents provoked by third parties. This solution appears particularly appropriate where compulsory auto insurances are discarded as in the state of New Hampshire.

6.3 Sources of Good and Evil

Along with the dramatic increase in the military sector over the past decade, robots have spread in both the industrial and service fields. Today, we deal with a number of robotic cleaners, surveyors, inspectors, entertainers, handicap assistants, space travellers, manufacturers of food and beverages, textile and leather producers, hunters, fishers, miners, farmers, doctors, nurses, scientists, academic and PR assistants. So far, legal systems have governed this panoply of robotic applications as they did with previous technological innovations. Rather than agents, let alone persons in the legal domain, robots have been regulated as sources of responsibility for designers, manufacturers, suppliers and users of such machines. The normative efforts of the law therefore have concerned liability for the behaviour of robots that may jeopardize foundational elements of society, or compensation for damages provoked by wrongdoing, in both criminal and civil law. In addition to the precautionary principle and the administrative powers of regulatory authorities, the way in which the law has tackled the challenges of robotics mostly revolves around the use of strict liability techniques. This is what we have ascertained through the various steps of the phenomenology of *Picciotto Roboto*, and in both the field of contracts and torts, where people are held responsible for damages or harm provoked by such machines, regardless of any illicit or culpable behaviour. The traditional legal viewpoint thus attributes to robots the dangerous propensities of animals and children, or, conversely, risky activities and potentially hazardous sources of harm.

In light of strict liability norms, as the default rule of today's legal systems, a crucial exception is represented by clauses of immunity in criminal law. Here, the golden rule allows individuals to evade responsibility in the name of the principle of legality and the rule of law. Even though most robotic crimes represent traditional types of offenses committed through such applications as innocent means in the hands of a human's *mens rea*, I have insisted throughout this work on a parallel with the new generation of computer crimes first introduced in the 1990s. It is indeed likely that robots will produce a novel generation of loopholes in the criminal law field, forcing lawmakers to intervene at both national and international levels. Still, such a condition of immunity, following from the principle of legality, is at times legitimate. The focus in Sect. 3.3 was on international agreements on the laws of war as well as humanitarian and human rights law. Attention was drawn in Sect. 3.5 to constitutional norms and statutory rights. In both cases, the aim was to stress that law enforcement officers, political authorities and military commanders are generally protected as long as the use of robots does not breach basic norms of the system, *e.g.*, constitutional safeguards.

This traditional framework for robots as sources of damages and individual responsibility for other agents in the system, gives rise to three concerns. First, regarding today's policies of strict liability, they present several drawbacks that suggest adopting policy changes, such as forms of negligence-based liability or the direct accountability of robots examined in the previous section. Clauses of no-fault liability might allocate risk and responsibility inefficiently, as in many cases of malfunction error, where third parties are the least-cost avoider of the risk. Moreover, strict liability rules might prevent people from producing and using a number of fruitful applications such as service robots for domestic and personal use. New technologies tend to be dangerous and, therefore, strict liability rules often represent the proper technique to scale back such kind of activities

(Posner 2007). However, some of these rules, when applied to the field of robotics, seem more a relic, than a rational outcome, of cost-benefit analysis.[11]

Secondly, concerning today's clauses of criminal immunity, we should distinguish between new crimes that humans can sometimes be charged with if they damage or destroy their robots, and cases concerning today's clauses of immunity, in such fields as the laws of war and international humanitarian law. As mentioned above in Sect. 3.3.3, what makes the use of robot soldiers critical depends on the technical difficulty of designing these machines so as to enable them to distinguish between friends and foes, and abiding by principles of military conduct like proportionate use of force or discrimination between soldiers and civilians. Similarly to previous technological advancements in chemical, biological or nuclear weapons, an international agreement is thus urgently needed, since analogy is inadequate to determine whether all types of autonomous weapons should be considered unlawful as such. Whereas both the UN General Assembly and its Secretary-General Ban Ki-Moon have been quiescent up to the date of publication of this book, it is noteworthy that the condition of immunity for the use of robot soldiers goes hand in hand with no-fault responsibility for the employment of both industrial and service robots in the civil sector. From this outlook, current clauses of criminal immunity look more like a matter of privilege, than sound protection from arbitrariness.

Thirdly, we may follow the advocates of the front of robotic liberation who suggest charging humans for abuses of their machines. Analogy may in fact fall short in likening the protection of such machines to current sanctions in cases of vandalism, intentional misuse of power, etc. Yet, *pace* the front of robotic liberation, the principle should also apply the other way around, so that once the use of this technology is deemed as illegal, robots can meaningfully represent a target of human censorship, *e.g.*, monitoring and modification, removal or deletion without backup.[12] Whilst these punitive sanctions do not directly involve the owner of the robot, they nonetheless affect the owner as well, since robots will increasingly raise psychological issues related to their interactions with humans as a matter of learning and adaptation. Recall the parallel with children and animals, as stressed above in Sect. 5.2, so that, in the case of robots for personal and domestic use, humans have to satisfy the social drives of the machine by responding to its internal needs. At times, the lawful removal or annihilation of such robots will be even worse than today's "three strikes" doctrine in the field of

[11] See above in Sects. 4.3.2 and 5.4.2.

[12] See above in Sect. 2.3.1.

Table 6.3 The challenges of today's laws of robots as a source of damage

Robot behaviour	Responsibility	Immunity	Strict liability	Unjust damages
As legal person	In all the fields	Sci-Fi scenarios	Sci-Fi scenarios	Sci-Fi scenarios
As strict agent	Contracts, Torts	Borderline	Why not?	Why not?
Source of harm	**In all the fields**	**Innovation**	**Status Quo**	**Innovation**

computer crimes. In this latter case, as a part of the graduated system which ends up with user disconnection after three warnings of copyright infringement, humans are temporarily banished from the internet. In the case of robots, monitoring, modification, removal or deletion of some robots for personal or domestic use bring us back to the words of Dostoevsky quoted at the beginning of Chap. 3: "If the human has a conscience he will suffer for his [*i.e.*, the robot's] mistake. That will be punishment as well as the prison."

Dealing with robots as a source of damages and responsibility for other agents in the system, the challenges of today's laws of robots can be summed up with a final table. This complements the Sci-Fi scenarios of Table 6.1 and the problems brought on by machines considered as proper agents in the fields of contracts and torts of Table 6.2. Let us have a look at the final row of Table 6.3:

In a nutshell, the distinction between traditional approaches and the need for new robotic policies suggests that we should retain the current rules of no-fault liability as the pillar of the system and, yet, today's legal framework should be amended by both curtailing cases of criminal immunity and inserting new clauses of negligence-based liability. This idea converges with the conclusion of the previous section, in that no new immunity policies are required in the civil (as opposed to the criminal) law field, but new mechanisms of responsibility appear necessary for both contractual and extra-contractual obligations. This means that four out of nine cases of responsibility for the behaviour of robots should be judged under a strain, that is:

(a) Immunity for humans bearing responsibility for the care of robots and their behaviour (I-3 of Table 1.1);
(b) Strict liability of robots conceived as proper agents in the field of contracts (SL-2 of Table 1.1);
(c) Unjust damages concerning robots as contractual agents (UD-2 of Table 1.1); and
(d) Unjust damages related to robots as a source of responsibility for other agents in the legal system (UD-3 of Table 1.1).

More particularly, in light of specific differences between criminal law, contracts and torts, 8 out of 27 possible cases should be under scrutiny; namely,

I-3 for criminal immunity of political authorities, military commanders, and a new generation of robotic crimes, SL-2 and UD-2 for both contractual and extra-contractual robotic obligations, and UD-3 for humans in all the fields of the law. Let us examine these cases separately in the next section.

6.4 Levels of Complexity

Complexity is a complex notion of its own. At the conference in complex engineering, co-sponsored by MIT and the Santa Fe Institute in 1999, Seth Lloyd pinpointed *31 Measures of Complexity*, to describe, reproduce and determine the degree of organization of what can be deemed as complex as a bacterium or an investment scheme. Two years later, when the paper was published in the "IEEE Control System Magazine" (1999, 2001), the measures of complexity increased to 42. In this context, it suffices to rely on Gregory Chaitin's notion of complexity in terms of information, so as to shed light on three different aspects of the aim of the law establishing the conditions of legitimacy for technological development and innovation. Dealing with the law as a meta-technology, the phenomenon will become all the more complex as the quantity of information grows and its theoretical compression decreases (Chaitin 2005). Once the complexity of the law in terms of informational compression is grasped, it can be fruitful to examine today's debate on the simplification of the law and three different ways by which the law can be understood in terms of information. The aim of this section is to determine how the complexity of the subject matter that today's legal systems aim to govern, namely, robotics technology, affects the complexity of the law. These three different levels of complexity are illustrated with Fig. 6.2:

First, the formula "complexity of the law" can be grasped as opposite to that which is simple, and even charged with ideological intent. In *Simple Rules for a Complex World* (1995), for example, Richard Epstein claims that the "complexity of legal rules tends to place the power of decision in the hands of other people who lack the necessary information and whose own self-interest leads them to use the information that they have in socially destructive ways." In contrast to simplification, even public organizations and institutions often refer to that which is complex, to stress the evil effects of intricate law-making in terms of anxiety and panic. Such a view, especially popular in France, is summed up by the remarks of the *Conseil d'État* in its Public Report (2006) focused on law, complexity and globalization: "la complexité croissante de notre droit est devenue une source majeure de fragilité pour notre société et notre économie."

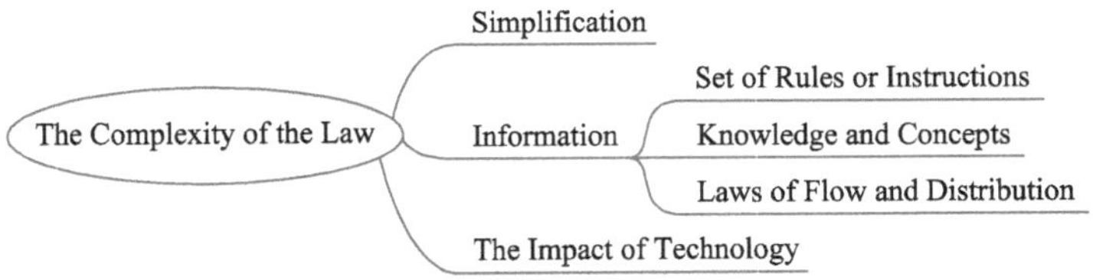

Fig. 6.2 Levels of legal complexity in the governance of robotics

These arguments are important for the accessibility of the law and, hence, the principle of legality is at risk when the intricacy of the regulations ends up in a legal labyrinth. Think about section 67(1) of the UK Rating & Valuation Act (1925) as a prototype of this kind of complex and destructive regulation: "If any difficulty arises in connection with the application of this Act *to any exceptional area*, or the preparation of the first valuation list for any area… the Minister [of Health] may by order remove the difficulty, or constitute any assessment committee, or declare any assessment committee to be duly constituted, or *do any other thing which appears to him necessary or expedient* for securing the preparation of the list."[13] Significantly, on 24 March 1988, the Italian Constitutional Court declared Article 5 of the Italian criminal code partially void, ruling that ignorance of the law constitutes an excuse for the citizen when the law is formulated in such a way that leads to obscure and contradictory results (sentence no. 364/88).

However, it does not follow, *pace* Epstein, from the pathology of the law that simple rules guarantee the transparency of the system. Lawmakers can endorse simple provisions and still end up with a limited predictability of the evolution of the law as well as a partial knowledge of the dynamic connections between the components of the system. Rather than synonymous with labyrinthine and astonishing intricacy, complexity may refer to the properties of a multi-agent system that adapts to the environment through learning and evolutionary processes, such as sophisticated signalling and information mechanisms. These systems are characterized by a collective behaviour that emerges from large networks of individual components, although no central control or simple rules of operation direct them. Current work on artificial intelligence and the complexity of the law makes this point clear (Casanovas et al. 2010), in connection with such fields as network theory, legal knowledge management, information and negotiation systems, ontologies in the legal domain, software agent systems,

[13] In Bingham (2011: 48–49), italics added.

and more. Three fundamental aspects of the law as a meta-technology are highlighted by considering the complexity of the law in terms of information and *vice versa*:

(a) The normative complexity of the law as a set of rules or instructions for the determination of other informational objects;
(b) The knowledge and concepts framing the function and representation of a shared legal terminology; and
(c) The laws of distribution of legal information hinging on the statistical properties of such quantities as the edges and diameters of the network.

The overall idea is that complexity does not necessarily entail uncertainty or legal chaos. Rather, according to the seminal remarks of *Law, Legislation and Liberty*, complexity is the key to understanding the very differences between deliberate human arrangements and the emergence of spontaneous orders (Hayek 1982). We return to this below.

The final step of the analysis on the complexity of the law should be on that which the law aims to govern, namely, the complexity of robotics technology. The set of concepts and ways of legal reasoning setting the conditions of legitimacy for the design, construction and use of robots should be further examined in the light of how technology impacts on legal know-how. Three hypotheticals have been further assessed in this book:

(a) Cases where the advancement of robotic technology does not seem to affect the principles and rules of today's legal systems, *e.g.*, I-1, SL-1, UD-1 and some SL-3s of Table 1.1;
(b) Cases where robotic technology impacts on pillars of current legal frameworks and, still, the use of analogy as well as the principles of the system allow lawyers to provide unequivocal solutions, *e.g.*, I-2 and some SL-3s and UD-3s of Table 1.1;
(c) Cases where there is no "general agreement in judgments as to the applicability of the classifying terms" (Hart 1994: 123). Such hard cases were stressed with the hypotheticals I-3, SL-2, UD-2 and some UD-3s of Table 1.1. At times, political decisions, rather than legal expertise, are crucially at stake in this context.

The focus next in Sect. 6.4.1 is on the differences between the traditional viewpoint that the law is a means of social control, and the level of abstraction adopted in this book on the aim of the law to govern technological advancement and innovation. By taking into account the impact of robotic technology on today's legal systems, the different levels of complexity concerned by this impact are more deeply illustrated in Sect. 6.4.2.

6.4.1 Technologies of Social Control

I have insisted on the formula "law as meta-technology" to set the proper level of abstraction, namely the set of features or observables of the analysis that concern the aim of the law to govern technology. Rather than dealing with the essence of the legal phenomenon, the analysis has dwelt on the complex set of powers, principles and provisions of the system, with which the law determines the conditions of legitimacy and responsibility for the design, construction, supply, and use of technological artefacts. From this point of view, special attention was drawn to the conditions in which individuals find themselves confronted with issues of legal responsibility, and the different ways robotic applications can be grasped as persons, proper agents, or sources of responsibility for other agents in the system. In connection with clauses of immunity, strict liability, and fault-based responsibility, this level of abstraction discerned 27 observables concerning the behaviour of robotic applications and, moreover, specific cases that should be judged under a strain in the fields of crimes, contracts and torts. Returning to the notions of causation and formal accountability summed up with the formula "if A, then B," the normative outcomes of the system seemed ultimately hard, or problematic, in cases I-3, SL-2 and UD-2 & 3 of Table 1.1.

This stance on the law as meta-technology, however, should not be confused with the idea of the law as made of commands enforced through physical sanctions (Kelsen 1945/1949). By examining the legal consequences (B) that follow from the hypotheses of harm or damages provoked by robotic applications (A), the set of powers, principles and provisions, with which the law aims to set the conditions of legitimacy for technological advancement and innovation are only a part, albeit a crucial one, of the legal order. Going back to the different approaches to the idea of complexity mentioned in the previous section, consider chapter 2 of the first volume of *Law, Legislation, and Liberty* (1973), where Hayek differentiates the regulative efforts of lawmakers, or *taxis*, from the law as both an evolutionary process and a spontaneous order, *i.e.*, that which Hayek identified with the idea of *kosmos*. This distinction is critical since the information lawmakers would need to direct the evolutionary process of the law, *e.g.*, the laws of robots, far exceeds the capability of any political planning. In the words of Hayek, "one of our main contentions will be that very complex orders, comprising more particular facts than any brain could ascertain or manipulate, can be brought about only through forces inducing the formation of spontaneous orders" (*op. cit.*, 38). This twofold level of complexity was assessed in Sect. 5.4.2, where the precautionary principle and its obverse, openness, were under scrutiny. In that context, the irreducibility of *kosmos* to *taxis* in the law was

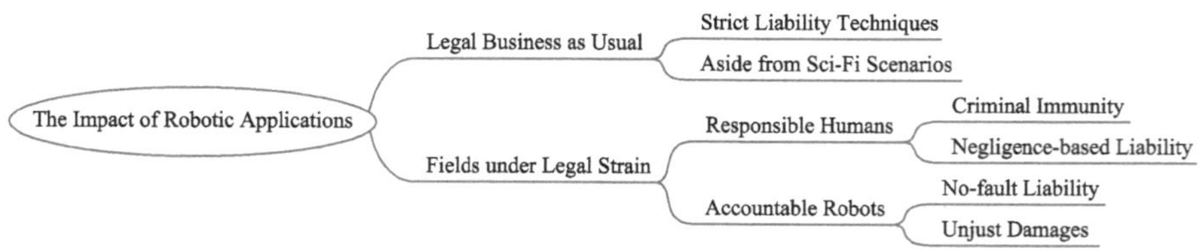

Fig. 6.3 Four robotic challenges to law as meta-technology

illustrated with the U.S. Supreme Court's ruling in the CDA case from June 1997, so as to suggest a strong rationale for engaging in scientific research and developing technological applications.[14] As the default rule, the burden of proof should fall on those who reckon that a certain technology is incapable of lawful uses or that the threats and risks outweigh potential benefits. The current debate on the legitimacy of employing robot soldiers in battle revolves around this allocation of the burden of proof.

Focusing on how legal systems establish the conditions of legitimacy and responsibility for the development of technology, a further reason why the formula "law as meta-technology" cannot be grasped as a variant of the idea that the law is a method, or technique, for social control, has to be stressed. Even though the law can affect technological advancement and innovation, technology also impacts on principles and pillars of the law: so far, we have seen that 8 out of 27 cases of responsibility for the behaviour of robots need further investigation. By taking into account specific differences and similitudes between criminal law, contracts and torts, such cases turned out to be four out of nine: see above in Sect. 6.3 and Table 6.3. Let aside Sci-Fi scenarios and "legal business as usual," namely cases where the advancement of robotic technology does not seem to affect principles and rules in today's legal systems, Fig. 6.3 illustrates which cases of the law are under stress:

First, let us look at clauses of criminal immunity. Besides the employment of robot soldiers in warfare, amendments to clauses of criminal immunity concern new types of crimes through robotic *actus rei* and even prosecution against humans for crimes committed against their machines. These issues were discussed above in Sect. 3.4.1 and in the previous section.

Second, the endorsement of strict liability rules does not follow from the need to amend some of the current clauses of immunity. Rather than no-fault responsibility, new provisions for negligence-based responsibility and other

[14] See above Sect. 5.4.1.

types of personal fault for humans appear necessary in all fields of the law as seen above in Sects. 3.5, 4.3.2 and 5.2.2.

Third, cases of negligence-based liability in tort law suggest distinguishing between plain cases raised by robots-as-means of human industry from some hard cases induced by robots-as-agents in the civil (as opposed to the criminal) law. A number of schemes for the accountability of robots concerning rights and obligations of their own contracts, such as the digital *peculium*, registers and insurance models, were examined in Sects. 4.4.1 and 4.5.1. Together with amendments to current clauses of criminal immunity, this is the field where the intervention of lawmakers is most urgent.

Fourth, clauses of strict liability and negligence-based responsibility for the behaviour of robots make sense in the field of torts, because the hypothetical of robots damaging third parties outside their working activities is problematic as raised above in Sects. 5.3 and 6.3. Moreover, such cases should be considered in connection with a panoply of robotic applications for domestic and personal use, such as robo-toys and robo-nannies. The need for new robotic policies has been stressed above in Sects. 5.4 and 6.2 dealing with the unjust damages provoked by such machines in the field of torts.

A final distinction however is necessary, that is, between cases where analogy as well as other principles of interpretation allows lawyers to provide solutions, and cases where political decisions, rather than legal expertise, are required. This distinction brings us back to the debate on whether and how the existence and content of the law can always be determined on the basis of its own sources. The different impact that the autonomy of robots has on the law finally reverberates on the ways we should grasp the hard cases raised by technology.

6.4.2 The Political Requirement

The analysis of the different levels of complexity invoked by the impact of robotic technology on the law has been summed up with the legal observables of Table 6.3 and Fig. 6.3. Needless to say, the intricacy of the analysis and, hence, the complexity of the model can be enhanced, by adding further fields to the study of robotic crimes, contracts and torts, *e.g.*, administrative law and the legal responsibility of regulatory authorities granting certificates to, say, civil UAVs as seen above in Sect. 5.4.1. However, the observables suffice to distinguish plain from hard cases in which the applicability of the classifying terms sparks general disagreement. This is occurring with some clauses of criminal immunity and negligence in criminal law and torts, unreasonable conduct of robotic agents in tort law, and accountable robo-traders for their business and agreements. How should lawyers deal with such hard cases?

Some, as Herbert Hart in *The Concept of Law* (1961: 128), reckon "there is no possibility of treating the question raised by the various cases as if there were one uniquely correct answer to be found, as distinct from an answer which is a reasonable compromise between many conflicting interests." Others have proposed a solution hinging on the principles of the system, conceived as normative statements with a deontological, rather than teleological, meaning. By following the logic of yes or no, or what is good for all, Ronald Dworkin endorses this idea, claiming that a "right answer" can be found for every case at hand. Jurists should identify the principles of the system that fit with the established law, so as to apply such principles in a way that interprets the case in the best possible light. As Dworkin states in *A Matter of Principle* (1985), this effort emphasizes the parallel between the law and literature, as stressed above in Sect. 2.1.1, because we "must read through what other judges in the past have written not only to discover what these judges have said, or their state of mind when they said it, but to reach an opinion about what these judges have collectively done, in the way that each of our novelists formed an opinion about the collective novel so far written" (*op. cit.*, 159). Although "some critics, including Brian Barry and Joseph Raz, suggest that I have changed my mind about the character and importance of the one-right-answer claim," Dworkin retrospectively claims, in *Justice in Robes* (2006: 266), "for better or for worse, I have not."

Whether or not Dworkin changed his mind, there are circumstances in which general disagreement depends on the fact that there are many right answers out there. Whereas analogy and the principled legal reasoning at times provide unequivocal solutions, a number of issues often remains open, as cases of criminal negligence, agenthood in contracts and policies of tort law illustrate in the laws of robots. Solutions vary in different traditions, customs and legal cultures, as the comparison between the American and the Italian models have shown in the field of tort liability, *e.g.*, forms of negligence-based accountability vs. traditional policies of no-fault responsibility. This is what *Law's Empire* seems to suggest after all: "For every route that Hercules took from that general conception to a particular verdict, another lawyer or judge who began in the same conception would find a different route and end in a different place, as several of the judges in our sample cases did. He would end differently because he would take leave of Hercules, following his own lights, at some branching point sooner or later in the argument" (Dworkin 1986: 412).

There is nonetheless a set of further cases in which general disagreement depends more on different moral and political assumptions than technicalities of legal expertise. In addition to "the fundamental question of whether lethal force should ever be permitted to be fully automated," according to the phrasing of Christof Heyns' 2010 Report to the UN General Assembly, think of whether and to what extent new robotic offences should be established. This was the option taken by international lawmakers with the Budapest Convention on Cybercrime in November 2001. In light of the current debate on whether a certain type of drone design should be considered legal in the field of robotics military technology and, moreover, what should be the design of the new environment of such a human-robot interaction, a reasonable compromise on the basis of legal expertise, rather than the search for any right answer, is at stake.

Admittedly, some types of robots, such as NAO or HRP-4C, are as lovable as the design of the TeleFunken U-47 praised by Zappa in *Joe's Garage* and, yet, many political decisions have to be taken in a world crowded by robots and other artificial agents. This has already occurred in traditional fields, where the focus is not only on the responsibility of the game players, but of the game designers as well. The legal design of this new environment has been problematic so far, in such fields as data protection law, copyright and computer crimes. Just reflect on today's debate on the filtering of the web, transparency of smart environments, protection of personal data and intellectual property, freedom on the internet of the things and surveillance through ambient intelligence. The ways lawmakers shape the environment of online human interaction necessarily reverberate on how humans may interact with their robots. How the law establishes the conditions of legitimacy for the production and use of technology (*i.e.*, Kelsen's A), so as to determine who is responsible when something goes wrong ("B"), is just as important as how the environment of the new human-robot interaction looks like. The conclusions of the book address this final issue.

Conclusions

The broader one's understanding of the human experience, the better design we will have.

Steve Jobs, The Next Insanely Great Thing (Wired, February 1996)

The "laws of robots" can be interpreted in a twofold way according to how the genitive case of the formula is understood, in either an objective or subjective manner. Grasping it as an objective genitive, the formula reminds us of the traditional viewpoint that considers robots the subjects of legal regulations establishing the conditions for human liability as to the damages or harms provoked by such machines. As a subjective genitive, *vice versa*, the formula stresses that which is specific of robots as the authors of the activity governed by the law. Aside from the front of robotic liberation, and claims as to the full-fledged personality of these machines, we have seen circumstances where a restricted personhood of robots makes sense for pragmatic reasons in the civil (as opposed to the criminal) side of the law. New forms of accountability for the behaviour of robots can strike a balance between the parties to a contract, or in the field of extra-contractual obligations. In addition to the objective or subjective uses of the genitive in the formula "the laws of robots," focus was on humans and machines conceived of as game players in the legal framework. By examining the conditions of responsibility in human-robot interaction, 27 hypotheticals were analysed under criminal law, contracts and torts. The aim was to pinpoint the cases of the laws of robots which are under a strain.

Still, the aim of the law to govern technology does not only have to do with agents in the legal field, since this aim concerns the provisions and norms that are shaping the environment of human-robot interaction as well.

U. Pagallo, *The Laws of Robots: Crimes, Contracts, and Torts*, Law, Governance and Technology Series 10, DOI 10.1007/978-94-007-6564-1,

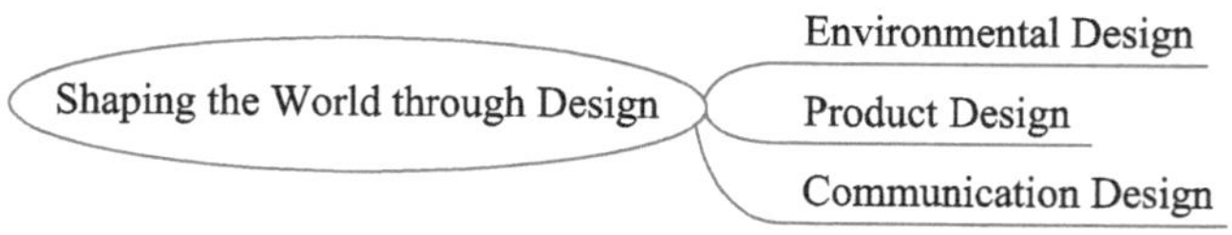

Fig. A.1 Three roads to design

This distinction, between game players and game designers, is not novel in the legal field. Reflect on traditional forms of enforcement, such as the installation of speed bumps in roads as a means to reduce the velocity of cars (lest drivers opt to destroy their own vehicles). Moreover, the current information revolution has forced legal systems to resort to more sophisticated ways of enforcement through the design of products and processes, much as the structure of spaces and places. Whereas, in *Code and Other Laws of Cyberspace* (1999), Lawrence Lessig lamented the lack of research involving the impact of design on both social relationships and the functioning of legal systems, it is noteworthy that this gap has been filled in just a few years. Think of work on privacy, universal usability, informed consent, crime control, self-enforcement technologies and more.[1] Not surprisingly, there is a variety of design approaches today, such as in data protection law (*e.g.*, privacy by design), copyright (*e.g.*, filtering systems as those established by the UK Digital Economy Act, or "DEA," from 2010), computer crimes (*e.g.*, information security systems against cyber-attacks), and so forth. While these mechanisms aim to frame the environment of current online interaction, they also concern how a world crowded by robots and artificial agents can be designed. By following the seminal remarks of Norman Potter in *What is a Designer* (1968, new ed. 2002), three different ways of conceiving the notion of design so as to work out the forms of our world, should be distinguished, namely, designing spaces (environmental design), objects (product design) and messages (communication design). These different aspects of design are illustrated with Fig. A.1:

As an illustration of the first kind of design, think about people's anonymity and the issue of protecting their privacy in public. While the use, say, of close circuit televisions, or "CCTVs," proliferates and seems unstoppable, it is feasible to design video surveillance systems in public transportation networks in such a way that the faces of individuals are not recognizable. That which the European authorities on data protection proposed in their document on *The Future of Privacy* (WP29 2009), can be extended to the video cameras of civilian drones, as seen above in Sects. 3.4.1 and 4.5.

[1] See, for example, Shneiderman (2000), Friedman et al. (2002), Katyal (2002, 2003), Borning et al. (2004), and Zittrain (2007).

The second kind of design has to do with the ways products can influence the behaviour of their users and the protection of their rights. Consider cases where making personal data anonymous is conceived of as a priority, so that matters of design involve how to organize data processes and products. A typical instance is given by the processing of patient names in hospitals via information systems: here, patient names should be kept separate from data on medical treatments, or health status, through, *e.g.*, the use of smart cards. From a legal viewpoint, issues of design arise in relation to defective products and if the lamented defect, as the proximate cause of the injuries suffered by the plaintiff, appeared while the product was under the manufacturer's control. These were issues of product design at stake in *Mracek v. Bryn Mawr Hospital* with the malfunctioning of a da Vinci robot.

Finally, as an example of communication design, think of the public complaints against Facebook's data protection policies. Some years ago, the social network announced on 26 May 2010 that it had "drastically simplified and improved its privacy controls," which previously amounted to 170 different options under 50 data protection-related settings. Regardless of whether the default configuration of Facebook has effectively been set to record only the name, profile, gender and networks of the user, that which is important to stress here is how interaction and communication depend on the design of the interfaces. In the case of Facebook, "friends" should no longer be automatically included in the flow of information, whilst users could finally turn off platform applications, such as games, widgets, and the like. In the case of robots, an example of communication design is given by the HRI work on the caretaker paradigm as examined above in Sect. 5.2. According to this robot-centred approach, the aim is to design robots with emotional and social needs to which humans can respond.

A further distinction has to do with the subjects of design, *i.e.*, places, products and organisms. This latter case concerns plants grown through OGM technology, genetically modified animals such as Norwegian salmons, or the current debate on human, post-humans, and cyborgs. Such engineering has been considered above in Sects. 5.4.1 and 6.5.1. On one hand, legal systems increasingly approach risks and threats of highly sensitive technologies with the precautionary principle. On the other hand, we discussed the thesis of the front of robotic liberation with the similarities between the combination of our biological design and social conditioning, and the programming of some smart robots (Chopra and White 2011: 176).

In this context, another aspect of design is particularly relevant, namely, the different goals according to which the environment of human-robot interaction can be framed. By embedding legal constraints in technology, the aim can alternatively be to encourage change in social behaviour,

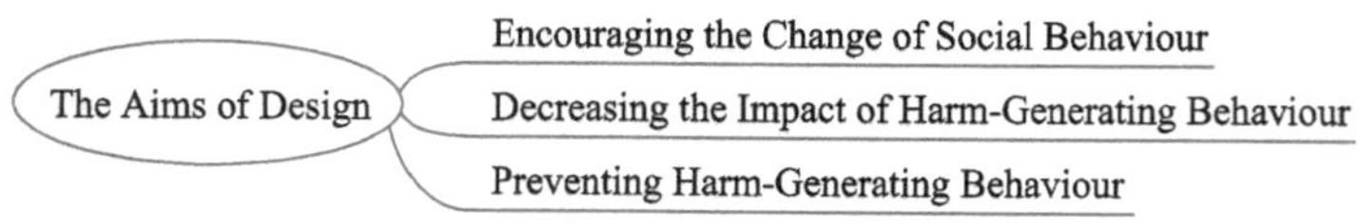

Fig. A.2 A teleological approach to design

to decrease the impact of harm-generating conduct or to prevent it even from occurring. Figure A.2 sums up this further approach to design:

As an illustration of the first goal of design, consider the example of robo-traders and how engineers intend to map their behaviour through incentives based on trust (*e.g.*, reputation mechanisms) or trade (*e.g.*, services in return). Design may also encourage change in conduct by widening the range of options available via user-friendly interfaces or transparent setting options. This is what occurs with modifications to interfaces that increase, or reduce, the prominence of a default setting, so as to allow users to configure and use their software as they deem appropriate.

As an example of the second modality of design, think about security measures. Here, the aim is not to encourage or induce humans, and robots, to change their behaviour, *e.g.*, the installation of speed bumps in roads as a means to reduce the velocity of AI chauffeurs. Rather, think of air-bags that reduce the impact of harm-generating conducts. Such mechanisms can be proactive: for example, the default configuration of ICT interfaces can ensure that values of design are appropriate for novice users and, still, improve the system's efficiency.

The final aim of design is the most relevant one in this context. There are a number of cases, where both lawmakers and private companies intend to prevent social behaviour from occurring through the use of self-enforcing technologies. In the field of copyright and intellectual property, for instance, most of the efforts have focused on how to safeguard these exclusivity rights through the development of digital right management ("DRM") systems. By enabling right-holders to strictly regulate the use of their own copyright protected works, companies would prevent irresolvable problems concerning the enforceability of national norms and the conflicts of law at the international level. Significantly, in his *Thoughts on Music* (2007), Steve Jobs conceded that DRM compliant systems raise severe challenges of interoperability and, hence, antitrust issues. Moreover, individual behaviour would unilaterally be determined on the basis of technology, rather than by the choices of the relevant political institutions.

This kind of environmental design has been furthered by current policies on the internet. Some states, like China, have built up systems of filters and re-routers, detours and dead-ends, to keep individuals on the state-approved online path. Others, Western democracies and authoritarian regimes alike, have endorsed the "three strikes" doctrine as a part of a graduated system that ends up with a user internet disconnection after three warnings of alleged copyright infringements. Whilst in December 2010, some members of the EU Commission proposed adopting a system of filters in order to control the flow of online information, there are risks of paternalism as well, since some lawmakers want to protect citizens even against themselves. Consider some versions of the principle of "privacy by design" and the intention to automatically protect personal data in every ICT system as its default position. The idea is that privacy safeguards should be at work even before a single bit of information has even been collected (Cavoukian 2010). Still, such automatic control appears even more problematic than the use of DRM technology for the protection and enforcement of digital copyright, since data protection does not represent any automatic "zero-sum game" between options of access and control over information in digital environments. Indeed, personal choices play the main role when individuals modulate different levels of such access and control, depending on the context and its circumstances. Furthermore, there is the technical difficulty of applying to a machine concepts traditionally employed by lawyers, through the formalization of norms, rights or duties. This difficulty has been stressed time and again in this book when dealing with Asimov's novels in Chap. 2, current research in military robotics in Chap. 3, some types of robo-traders in Chap. 4, and so forth. As a matter of fact, normative safeguards are often highly context-dependent and raise significant problems when reducing the complexity of a system where concepts and relations are subject to evolution. To the best of my knowledge, it is still impossible to program software so as to prevent forms of harm generating-behaviour as simple as defamation: these constraints emphasize critical facets of design underlying the use of allegedly perfect self-enforcing technologies. Reflect on three aspects of the problem:

First, there is the risk of updating traditional forms of paternalism, in that individual's behaviour would unilaterally be determined on the basis of automatic techniques rather than by individual choices on levels of access and control over information: "the controls over access to content will not be controls that are ratified by courts; the controls over access to content will be controls that are coded by programmers" (Lessig 2004).

Second, attention should be given to the difficulties of achieving such total control. Doubts cast by "a rich body of scholarship concerning the theory and practice of 'traditional' rule-based regulation bear witness of the impossibility of designing regulatory standards in the form of legal rules that will hit their target with perfect accuracy" (Yeung 2007).

Third, specific design choices may result in conflicts between values and, *vice versa*, conflicts between values may impact on the features of design: "some technical artefacts bear directly and systematically on the realization, or suppression, of particular configurations of social, ethical, and political values" (Flanagan et al. 2008). Even though legal systems help us overcome a number of conflicts between values, it is likely that the use of self-enforcement technologies in such fields as data protection or copyright would make conflicts between values even worse. Consider the impact of specific design choices, such as the opt-in vs. opt-out diatribe over the setting for users of information systems.

In light of today's debate on how design affects online interaction, let us restrict the focus of this analysis and examine the role of design in the laws of robots. In addition to projects encouraging agents to change their conduct (*e.g.*, speed bumps), or decrease the impact of harm-generating behaviour (*e.g.*, air-bags), think of design for AI cars, which should be able to stop or limit their speed according to the inputs of the surrounding environment. Here, preventing harm-generating conduct from even occurring impacts on the security of the robotic system through the use of driver checking mechanisms and cruise control, blind spot monitoring and traffic sign recognition, pre-crash schemes and so forth. Such systems will increasingly be connected to a networked repository on the internet that allows robots to share the information required for object recognition, navigation and task completion in the real world. The environment of AI car behaviour is thus designed as a complex multi-agent system where maintenance and safety contractors, traffic operators and internet controllers, interact with autonomous or semi-autonomous machines in order to avoid collisions, communication interferences, environmental concerns, and more. By considering the intricacy of this system, some reckon that a failure of legal causation could emerge as a result (*e.g.*, Karnow 1996). Certain suggest that the best method of accident control should be to scale back the activity through strict liability policies (Posner 1973: 180). Others claim that social and technical transactions run by artificial agents should be brought back under human control (Teubner 2007: 21). Yet, sweeping generalizations hardly fit the laws of robots: whereas some robotic applications, such as autonomous lethal weapons and some types of robo-traders, truly challenge basic pillars of the law, this is not the case with respect to other applications like da Vinci robots, NAOs, HRP-4Cs, etc. Accordingly, the aim of this book has been to introduce laypersons to the plain cases of the laws of robots so as to distinguish these cases of general agreement from the hard cases induced by robo-soldiers, robo-traders or AI chauffeurs.

Traditionally presented as a matter of facts vs. values, description vs. prescription, such different levels of analysis were magisterially summed up by Max Weber's ideal of *Wertfreiheit*. In his phrasing, "the capacity to distinguish between empirical knowledge and value-judgments, and the fulfilment of the scientific duty to see the factual truth as well as the practical duty to stand up for our own ideals constitute the program to which we wish to adhere with ever increasing firmness" (Weber 1904, ed. 1949: 58). As to the descriptive aspect of this book, the aim has therefore been to show that a relatively strong consensus still exists on the rules that govern the design, production and use of robots as well as on the consequences in terms of legal responsibility. Despite the complexity of robotic applications and the design of their surrounding environment, jurists generally agree on how to deal with responsibility pursuant to the liability model in accomplice cases within criminal law (Chap. 3), responsibility that depends on the voluntary agreement between private persons in the civil law field (Chap. 4), or strict liability that hinges on the idea of dangerous activities in tort law (Chap. 5). In all of these cases, there is no such thing as a failure of legal causation that suggests bringing robots back under human control.

As to the value-judgements of this book, on the other hand, the analysis has involved two different steps. First, identifying cases where the applicability of the classifying terms sparks general disagreement: diachronically, this emerged with the analysis of robot soldiers in Sect. 3.3.4, crimes of negligence intertwined with matters of legal causation in Sect. 3.5, the contract problem in Sect. 4.3.2, strict liability policies in Sect. 5.3, and so forth. Such hard cases were summarized by Table 6.3 and Fig. 6.3 in Chap. 6: drawing on the different reasons why pillars of the law, such as principles, concepts and ways of legal reasoning, are under a strain, the second step of the analysis concerned whether one right answer could be at hand, whether legal systems are instead open to alternative solutions, or political decisions need to be taken via, say, international agreements. On this basis, let me here take sides in today's debate by stressing which of these hard cases should have priority.

First, the regulation of robot soldiers in battle should have top priority, because of their hazardous effects on the environment and the human race. Current principles and provisions of the laws of war, international humanitarian law and human rights agreements do not regulate critical issues such as whether lethal force can be fully automated, or what set of parameters and conditions should regulate the use of these machines, *e.g.*, the US Air Force's claim that its drones have the same right of humans to defend themselves with ammunition. A solution could be to design robots that can target only weapons or operate in particular situations. Moreover, monitoring and verification mechanisms should allow for a determination of the locus of political and military decisions that, otherwise, could be very difficult to detect because of the increasing complexity of network-centric operations and the miniaturization of lethal machines. Whilst some 40 countries are currently developing autonomous weapons and other types of robot soldiers, this is a paradigmatic case where there is no such a thing as one right answer but, rather, a reasonable compromise between many conflicting interests should be found. Just as previous international agreements have regulated technological advancements over the past decades in fields such as chemical, biological and nuclear weapons, landmines, and the like, a similar UN-sponsored agreement is urgently needed to define the conditions of legitimacy for the employment of robot soldiers.

Second, there is the set of hard cases concerning the accountability of robots in the law of contracts. Contrary to the traditional viewpoint of robots as mere tools of the principal, so that humans should automatically be bound by all the operations of the artificial agent, new liability policies have to be taken into account. Indeed, a number of cases have shown that third parties, rather than individuals bearing responsibility for the care of their agents, are in the best position to prevent harm or damages and, thus, such third parties are the least-cost avoider of the risk. Furthermore, it makes a lot of sense to conceive certain types of robots as proper agents in the field of contracts, since the legal agency of these machines makes it clear that humans do delegate crucial cognitive tasks to their robots. This solution not only renders irrelevant several drawbacks of the traditional viewpoint, such as whether a robot is acting within certain legal powers, who should be held liable for conferring such powers, or whether users and operators can expect to evade responsibility for possible malfunctions of the machine. What is more, the personal accountability of robots demonstrates a fruitful way of striking a balance between the different human interests involved, namely, between the interest of the counterparties of robots to safely transact or interact with them, and the claim of users and owners of robots not to be ruined by the growing autonomy and even unpredictability of their behaviour.

Third, the new mechanisms of personal accountability for robots as well as clauses of negligence-based responsibility could properly be extended to the fields of torts. It is crucial here to distinguish the different types of robots humans will be dealing with in the foreseeable future. For example, personal accountability for the behaviour of robo-traders makes sense in tort law, because the hypothetical of robots damaging third parties outside their working activities appears problematic. *Vice versa*, clauses of negligence-based responsibility can replace some of today's strict liability rules in the aforementioned cases where third parties are the least-cost avoider of the risk. Still, dealing with service robots for domestic and personal use, most of the issues concerning responsibility for the behaviour of these machines are admittedly open. Legal systems could conceive robots in accordance with the responsibility of the American parent, so that defendants have to prove that their machine did not present any dangerous propensity or trait that is not typical of similar applications. Alternatively, according to the model of responsibility of the Italian parent in the field of extra-contractual obligations, defendants could avoid responsibility when evidence is given that they could not have prevented the harmful conduct of the robot, or that a fortuitous intervening event occurred. In any case, *pace* Dworkin, more than one right answer is possible.

Four, attention should be drawn to the use of robots as innocent means of human *mens rea*. In addition to crimes of war or against humanity, such as the Government of Sudan operating Iranian drones to assault civilians in the Nuba mountains of South Kordofan in April 2012, consider an increasing number of robots employed to physically alter US dollars, tiny drones employed in jewellery heists, or unmanned underwater vehicles used by Colombian drug traffickers. So far, such robotic *actus reus* can be prosecuted under current provisions of criminal law and still, no Sci-Fi imagination is necessary to envisage a new generation of robotic crimes that will force lawmakers to intervene, much as they did with a new generation of computer crimes in the early 1990s. Although what guise such a new robotic *actus reus* will assume is difficult to predict, we can imagine complex network-centric robotic applications that automatically collect and bring information to cloud servers, thereby replicating and spreading this data that could impinge on current privacy protections, copyright provisions, trade secrets and the like. Regardless of the specific content of such crimes, however, it is likely that such scenarios will concern the environmental design of human-robot interaction mentioned above.

One solution could be the use of self-enforcing technologies, much as those proposed in the case of robot soldiers, to prevent harm-generating conduct from even occurring. *Pace* the front of robotic liberation, none of the criticisms against such design policies, such as risks of paternalism and other ethical threats to individual autonomy, are in fact applicable to robots. Moreover, a number of Western lawmakers reckon that such measures, as automatic privacy by design and systems of filtering, are appropriate to impose order in online interaction. All in all, why should we not apply to tomorrow's robots the environmental design that some politicians propose for today's human interaction? Is it not a good thing to design robots in such a way as to prevent harm-generating conduct from occurring?

Definitely, there will be an increasing number of cases where such policies will be necessary for preventing robots from provoking accidental wars, economic meltdowns or traffic emergencies. However, aside from the technical difficulties in achieving such overall control, consider the set of service robots for domestic and personal use, such as the i-Jeeves 2.0 mentioned in

Chap. 4. Here, it is likely that the use of self-enforcing technologies would not only prevent robotic behaviour from occurring; such design policies may impinge on individual rights and freedom, by unilaterally determining how the artificial agents should act when collecting the information they need for human-robot interaction and tasks completion from networked repositories. This risk of modelling human behaviour through the design of their robots can be tackled with alternative design policies and new forms of legal accountability, such as the digital *peculium*. Likewise, security measures, such as user-friendly setting options or default mechanisms for the configuration of ICT interfaces, can ensure that values of design are appropriate for novice users, although allowing the robot to improve its own efficiency. Whereas further examples of design show how the use of self-enforcing technologies is not always necessary and, at times, can even be pernicious, let us therefore avoid conclusive generalizations. Law can govern technology through regulations and provisions that shape the environment of human-robot interaction without falling back on self-enforcing technologies. If there is no need to humanize our robotic applications, we should not robotize human life either.

References

Allen, Tom, and Robin Widdison. 1996. Can computers make contracts? *Harvard Journal of Law & Technology* 9(1): 26–52.

Allen, Colin, Gary Varner, and Jason Zinser. 2000. Prolegomena to any future artificial moral agent. *Journal of Experimental and Theoretical Artificial Intelligence* 12: 251–261.

Alston, Philip. 2010. *Report of the Special Rapporteur on extrajudicial, summary and arbitrary executions*. UN General Assembly, Human Rights Council, A/HRC/14/24/Add.6, 28 May.

Andonian, Sero, et al. 2008. Device failures associated with patient injuries during robot-assisted laparoscopic surgeries: A comprehensive review of FDA MAUDE database. *The Canadian Journal of Urology* 15(1): 3912–3916.

Andrade, Francisco, Paulo Novais, José Machado, and José Neves. 2007. Contracting agents: Legal personality and representation. *Artificial Intelligence and Law* 15: 357–373.

Aristotle. 1984. *Metaphysics*. Trans. W.D. Ross. In *The complete works of Aristotle*, ed. J. Barnes, vol. 2, 155-2-1728. Princeton: Princeton University Press.

Arkin, Ronald C. 2007. *Governing lethal behaviour: Embedding ethics in a hybrid deliberative/hybrid robot architecture*, Report GIT-GVU-07-11, Georgia Institute of Technology's GVU Center, Atlanta, GA.

Asaro, Peter. 2008. How just could a robot war be? *Frontiers in Artificial Intelligence and Applications* 75: 50–64.

Asimov, Isaac. 1985. *Robots and empire*. New York: Doubleday.

Asimov, Isaac. 1995. *The complete robot: The definitive collection of robot stories*. London: Harper Collins.

Barfield, Woodrow. 2005. Issues of law for software agents within virtual environments. *Presence* 14(6): 741–748.

Barrio, Fernando. 2008. Autonomous robots and the law. *Society for Computers and Law*. Retrieved from http://www.scl.org/site.aspx?i=ho0.

Bartneck, Christoph, Juliane Reichenbach, and Julie Carpenter. 2006. Use of praise and punishment in human-robot collaborative teams. In *Proceedings of the RO-MAN 2006 – The 15th IEEE international symposium on robot and human interactive communication*, Hatfield.

U. Pagallo, *The Laws of Robots: Crimes, Contracts, and Torts*, Law, Governance and Technology Series 10, DOI 10.1007/978-94-007-6564-1,

Bartolus de Saxoferrato. 1996. Digestum Novum. In *Commentaria*, vol. 6. Roma: Il Cigno, Galileo Galilei.

Beck, Ulrich. 1992. *Risk society: Towards a new modernity*. London: Sage.

Bekey, George A. 2005. *Autonomous robots: From biological inspiration to implementation and control*. Cambridge, MA/London: The MIT Press.

Bellia, Anthony J. 2001. Contracting with electronic agents. *Emory Law Journal* 50: 1047–1092.

Bingham, Tom. 2011. *The rule of law*. London: Penguin.

Borden, Lester S., Paul M. Kozlowski, Christopher R. Porter, and John M. Corman. 2007. Mechanical failure rate of Da Vinci robot system. *The Canadian Journal of Urology* 14(2): 3499–3501.

Borning, Alan, Batya Friedman, and Peter H. Kahn. 2004. Designing for human values in an urban simulation system: Value sensitive design and participatory design. In *Proceedings of eighth biennial participatory design conference*, 64–67. Toronto: ACM Press.

Breazeal, Cynthia. 2002. *Designing sociable robots*. Cambridge, MA: MIT Press.

Calude, Cristian (ed.). 2008. *Randomness and complexity. From Leibniz to Chaitin*. Singapore: World Scientific.

Canning, John S. 2008. Weaponized unmanned systems: A transformational warfighting opportunity, government roles in making it happens. In *American Society of Naval Engineers' (ASNE) Proceedings of Engineering the Total Ship (ETS) symposium*, Falls Church, VA.

Ĉapek, Karel. 1920. *Rossum's universal robots*. Trans. C. Novack. New York: Penguin (2004 edn).

Casanovas, Pompeu, Ugo Pagallo, Giovanni Sartor, and Gianmaria Ajani (eds.). 2010. *AI approaches to the complexity of legal systems. Complex systems, the semantic web, ontologies, argumentation, and dialogue*. Berlin/Heidelberg: Springer.

Castelfranchi, Cristiano, and Rino Falcone. 1998. Principles of trust for MAS: Cognitive anatomy, social importance, and quantification. In *Third international conference on multi-agent systems*. Paris, France: IEEE Computer Society.

Cavoukian, Ann. 2010. Privacy by design: The definitive workshop. *Identity in the Information Society* 3(2): 247–251.

Chaitin, Gregory. 2005. *Meta-math! The quest for Ω*. New York: Pantheon.

Chopra, Samir, and Laurence F. White. 2011. *A legal theory for autonomous artificial agents*. Ann Arbor: The University of Michigan Press.

Cicero. 1999. *On the commonwealth and on the laws*, ed. J.E.G. Zetzel. Cambridge: Cambridge University Press.

Clarke, Roger. 1993. Asimov's laws of robotics: Implications for information technology. *IEEE Computer* 26(12): 53–61.

Clarke, Roger. 1994. Asimov's laws of robotics: Implications for information technology. *IEEE Computer* 27(1): 57–66.

Comanducci, Paolo. 1986. Le tre leggi della robotica e l'insegnamento della filosofia del diritto. *Materiali per una storia della cultura giuridica* 36(1): 191–197.

Coudert, Allison P. 1995. *Leibniz and the Kabbalah*. Boston/London: Kluwer.

Croce, Benedetto. 1907. *Riduzione della filosofia del diritto alla filosofia dell'economia*. Bari: Laterza.

Datteri, Edoardo. 2011. Predicting the long-term effects of human-robot interaction. *Science and Engineering Ethics*, 29 July. (epub ahead of print.)

Dautenhahn, Kerstin. 2007. Socially intelligent robots: Dimensions of human-robot interaction. *Philosophical Transactions of the Royal Society B: Biological Sciences* 362(1480): 679–704.

Davis, Jim. 2011. The (common) laws of man over (civilian) vehicles unmanned. *Journal of Law, Information and Science* 21(2). doi:10.5778/JLIS.2011.21.Davis.1.

Dennett, Daniel. 1987. *The intentional stance*. Cambridge, MA: MIT Press.

Dennett, Daniel. 1997. When HAL kills, who's to blame? In *HAL's legacy: 2001's computer as dream and reality*, ed. D. Stork, 351–365. Cambridge, MA: MIT Press.

Diamond, Jared. 2005. *Collapse. How societies choose to fail or succeed*. London: Penguin.

Doorn, Neelke, and Sven Hansson. 2011. Should probabilistic design replace safety factors? *Philosophy and Technology* 24(2): 151–168.

Dworkin, Ronald. 1982. Law as interpretation. *Critical Inquiry* 9(1): 179–200.

Dworkin, Ronald. 1985. *A matter of principle*. Oxford: Oxford University Press.

Dworkin, Ronald. 1986. *Law's empire*. Cambridge, MA: Harvard University Press.

Dworkin, Ronald. 2006. *Justice in robes*. Oxford: Oxford University Press.

Elishakoff, Isaac. 2004. *Safety factors and reliability: Friends or foes?* Dordrecht/Boston/London: Kluwer.

Epstein, Richard Allen. 1995. *Simple rules for a complex world*. Cambridge, MA: Harvard University Press.

Epstein, Richard G. 1997. *The case of the killer robot*. New York: Wiley.

Ewald, William B. 1995. Comparative jurisprudence (I): What was it like to try a rat? *University of Pennsylvania Law Review* 143: 1889–2149.

Filmer, Robert, 1991. *Patriarcha and other writings*. Cambridge: Cambridge University Press.

Flanagan, Mary, Daniel C. Howe, and Helen Nissenbaum. 2008. Embodying values in technology: Theory and practice. In *Information technology and moral philosophy*, ed. J. van den Hoven and J. Weckert, 322–353. New York: Cambridge University Press.

Floridi, Luciano. 2007. Artificial companions and their philosophical challenges. *E-mentor* 5(22): 84–86.

Floridi, Luciano. 2008. The method of levels of abstraction. *Minds and Machines* 18(3): 303–329.

Floridi, Luciano. 2013. *Information ethics*. Oxford: Oxford University Press.

Floridi, Luciano, and Jeff Sanders. 2004. On the morality of artificial agents. *Minds and Machines* 14(3): 349–379.

Foster, Caroline. 2011. *Science and the precautionary principle in international courts and tribunals*. Cambridge: Cambridge University Press.

Franklin, Stan, and Art Graesser. 1997. Is it an agent, or just a program? A taxonomy for autonomous agents. In *Intelligent agents III. Proceedings of the third international workshop on agent theories, architectures, and languages*, ed. J.P. Müller, M.J. Wooldridge, and R. Nicholas, 21–35. Berlin: Springer.

Freitas Jr., Robert A. 1985. The legal rights of robotics. *Student Lawyer* 13: 54–56.

Friedman, Batya, Daniel Howe, and Edward Felten. 2002. Informed consent in the Mozilla browser: Implementing value-sensitive design. In *Proceedings of 35th annual Hawaii international conference on system sciences*, 247. Los Angels: IEEE Computer Society.

Gogarty, Brendan, and Meredith Hagger. 2008. The laws of man over vehicle unmanned: The legal response to robotic revolution on sea, land and air. *Journal of Law, Information and Science* 19: 73–145.

Goldberg, Ken, Eric Paulos, John Canny, Judith Donath, and Mark Pauline. 1996. Legal tender. In *ACM SIGGRAPH 96 visual proceedings, August 4–9*, 43–44. New York: ACM Press.

Gordley, James. 2006. *Foundations of private law: Property, tort, contract, unjust enrichment*. Oxford/New York: Oxford University Press.

Grodzinsky, Francis S., Keith A. Miller, and Marty J. Wolf. 2008. The ethics of designing artificial agents. *Ethics and Information Technology* 10: 115–121.

Habermas, Jürgen. 1996. *Between facts and norms*. Cambridge: Polity Press.

Hall, Storrs J. 2007. *Beyond AI: Creating the conscience of the machine*. New York: Prometheus.

Hallevy, Gabriel. 2011. Unmanned vehicles – Subordination to criminal law under the modern concept of criminal liability. *Journal of Law, Information, and Science* 21(2). doi:10.5778/JLIS.2011.21.Hallevy.1.

Hanson, Randall K. 1989. Parental liability. *Wisconsin Lawyer* 62: 24–28.

Hart, Herbert L.A. 1961. *The concept of law*. Oxford: Clarendon (2nd edn, 1994).

Hayek, Friedrich A. 1960. *The constitution of liberty*. Chicago: University of Chicago Press.

Hayek, Friedrich A. 1982. *Law, legislation and liberty: A new statement of the liberal principles of justice and political economy*. Chicago: Chicago University Press.

Hildebrandt, Mireille. 2010. *Criminal liability and 'smart' environments*. Conference on the philosophical foundations of criminal law at Rutgers-Newark, August 2009.

Hildebrandt, Mireille. 2011. *From Galatea 2.2 to Watson – And back?*. IVR world conference, August 2011

Hildebrandt, Mireille, Bert-Jaap Koops, and David-Olivier Jaquet-Chiffelle. 2010. Bridging the accountability gap: Rights for new entities in the information society? *Minnesota Journal of Law, Science & Technology* 11(2): 497–561.

Himma, Kenneth E. 2007. Artificial agency, consciousness, and the criteria for moral agency: What properties must an artificial agent have to be a moral agent? In *2007 Ethicomp proceedings*, 236–245. Tokyo: Global e-SCM Research Center & Meiji University.

Hobbes, Thomas. 1999. In *Leviathan*, ed. R. Tuck. Cambridge: Cambridge University Press.

HSC. 2007. *The sigma and delta scans*, research commissioned by the UK Office of Science and Innovation's Horizon Scanning Centre. *Foresight Annual Review 2007*, at 23.

JCSS. 2001. *Probabilistic mode code: Part 1—Basis of design*. Joint Committee on Structural Safety.

Jin, Linda X., Andrew M. Ibrahim, Naeem A. Newman, Danil V. Makarov, Peter J. Pronovost, and Martin A. Makary. 2011. Robotic surgery claims on United States Hospital websites. *Journal for Healthcare Quality* 11 (published online on 17 May).

Jobs, Steve. 2007. *Thoughts on music*. Retrieved at http://www.apple.com/hotnews/thoughtsonmusic/ on 22 Aug 2012.

Jonas, Hans. 1979. *The imperative of responsibility: In search of ethics for the technological age*. Chicago: University of Chicago Press.

Kahn, Peter H., Batya Friedman, Deanne R. Pérez-Granados, and Nathan G. Freier. 2006. Robotics pets in the lives of preschool children. *Interaction Studies* 7(3): 405–436.

Karnow, Curtis E.A. 1996. Liability for distributed artificial intelligence. *Berkeley Technology and Law Journal* 11: 147–183.

Katyal, Neal. 2002. Architecture as crime control. *Yale Law Journal* 111(5): 1039–1139.

Katyal, Neal. 2003. Digital architecture as crime control. *Yale Law Journal* 112(6): 101–129.

Kelly, Kevin. 2010. *What technology wants*. New York: Viking.

Kelsen, Hans. 1934/2002. *Pure theory of law*. Trans. B.L. Paulson and S.L. Paulson. Oxford: Clarendon.

Kelsen, Hans. 1945/1949. *General theory of the law and the state*. Trans. A. Wedberg. Cambridge, MA: Harvard University Press.

Kerr, Ian. 2001. Ensuring the success of contract formation in agent-mediated electronic commerce. *Electronic Commerce Research Journal* 1: 183–202.

Knight, Frank H. 1921. *Risk, uncertainty and profit* . Chicago: Chicago University Press. (reissue 2005 by Cosimo, New York.).

Krishnan, Armin. 2009. *Killer robots: Legality and ethicality of autonomous weapons*. Burlington-Surrey: Ashgate.

Krishnan, Armin. 2011. UVs, network-centric operations, and the challenge for arms control. *Journal of Law, Information, and Science* 21(2). doi:0.5778/JLIS.2011.21. Krishnan.1.

Kurzweil, Ray. 2005. *The singularity is near*. New York: Viking.

Latour, Bruno. 2005. *Reassembling the social: An introduction to actor-network-theory*. Oxford: Oxford University Press.

Lee, Seong Jae, Amy Greenwald, and Victor Naroditskiy. 2007. RoxyBot-06: An (SAA)2 TAC travel agent. In *IJCAI'07 proceedings of the 20th international joint conference on AI*, 1378–1383. San Francisco: Morgan Kaufmann.

Lerouge, Jean-François. 2000. The use of electronic agents questioned under contractual law: Suggested solutions on a European and American level. *The John Marshall Journal of Computer and Information Law* 18: 403.

Lessig, Lawrence. 1999. *Code and other laws of cyberspace*. New York: Basic Books.

Lessig, Lawrence. 2004. *Free culture: The nature and future of creativity*. New York: Penguin.

Levy, David. 2007. *Love and sex with robots: The evolution of human-robot relationships*. New York: Harper.

Lin, Patrick, George Bekey, and Keith Abney. 2008. *Autonomous military robotics: Risk, ethics, and design*. Report for US Department of Navy, Office of Naval Research. Ethics + Emerging Sciences Group at California Polytechnic State University, San Luis Obispo, CA.

Lloyd, Seth. 1999. *31 measures of complexity*. Complexity in engineering conference, co-sponsored by MIT and the Santa Fe Institute, 19–20 Nov, Cambridge, MA.

Lloyd, Seth. 2001. Measures of complexity: A nonexhaustive list. *IEEE Control Systems* 21(4): 7–8.

Locke, John. 1988. In *Two treatises of government*, ed. P. Laslett. Cambridge: Cambridge University Press.

Lolli, Gabriele, and Ugo Pagallo (eds.). 2008. *La complessità di Gödel*. Torino: Giappichelli.

Lorenz, Karl. 1971. Part and parcel in animal and human societies. In *Studies in animal and human behavior*, vol. 2, 115–195. Cambridge, MA: Harvard University Press. (first edition 1950.).

Luck, Michael, Peter McBurney, Onn Shehory, and Steven Willmott. 2005. *Agent technology: Computing as interaction*. AgentLink III, The European Coordination Action for Agent-Based Computing (IST-FP6-002006CA).

MacKie-Mason, Jeffrey K., and Michael P. Wellman. 2006. Automated markets and trading agents. In *Handbook of computational economics*, vol. 2, ed. Leigh Tesfatsion and L. Judd. Amsterdam: Elsevier. Available at SSRN: http://ssrn.com/abstract=974921.

McDaniels, Timothy, and Mitchell J. Small. 2004. *Risk analysis and society*. Cambridge: Cambridge University Press.

McFarland, David. 2008. *Guilty robots, happy dogs: The question of alien minds*. New York: Oxford University Press.

Michaelson, Greg, and Ruth Aylett. 2011. Special issue on social impact of AI: Killer robots or friendly fridges. *AI and Society* 26(4): 317–328.

Miller, Ross M. 2008. Don't let your robots grow up to be traders: Artificial intelligence, human intelligence, and asset-market bubbles. *Journal of Economic Behavior and Organization* 68(1): 153–166.

Moravec, Hans. 1999. *Robot: Mere machine to transcendent mind*. London: Oxford University Press.

Mosneron-Dupin, Fabrice, et al. 1997. Human-centered modeling in human reliability analysis: Some trends based on case studies. *Reliability Engineering and System Safety* 58(3): 249–274.

Nissenbaum, Helen. 2001. Securing trust online: Wisdom or oxymoron? *Boston University Law Review* 81: 101–131.

Pagallo, Ugo. 2010a. Robotrust and legal responsibility. *Knowledge, Technology & Policy* 23: 367–379.

Pagallo, Ugo. 2010b. The human master with a modern slave? Some remarks on robotics, ethics, and the law. In *The "backwards, forwards and sideways" changes of ICT*, ed. M. Arias-Oliva, T.W. Bynum, S. Rogerson, and T. Torres-Corona, 397–404. Tarragona: Universitat Rovira I Virgili.

Pagallo, Ugo. 2010c. As law goes by: Topology, ontology, evolution. In *AI approaches to the complexity of legal systems. Complex systems, the semantic web, ontologies, argumentation, and dialogue*, ed. P. Casanovas, U. Pagallo, G. Sartor, and G. Ajani, 12–26. Dordrecht: Springer.

Pagallo, Ugo. 2011a. The adventures of Picciotto Roboto: AI and ethics in criminal law. In *The social impact of social computing*, ed. A. Bissett, A. Light, A. Lauener, S. Rogerson, and T. Ward Bynum, 349–355. Sheffield: Sheffield Hallam University.

Pagallo, Ugo. 2011b. Killers, fridges, and slaves: A legal journey in robotics. *AI and Society* 26(4): 347–354.

Pagallo, Ugo. 2011c. Robots of just war: A legal perspective. *Philosophy and Technology* 24(3): 307–323.

Pagallo, Ugo. 2011d. Designing data protection safeguards ethically. *Information* 2(2): 247–265.

Pagallo, Ugo. 2011e. Guns, ships, and chauffeurs: The civilian use of UV technology and its impact on legal systems. *Journal of Law, Information and Science* 21(2). doi:10.5778/JLIS.2011.21.Pagallo.1.

Pagallo, Ugo. 2012a. Three roads to complexity, AI and the law of robots: On crimes, contracts, and torts. In *AI approaches to the complexity of legal systems. Models and ethical challenges for legal systems, legal language and legal ontologies, argumentation and software agents*, ed. M. Palmirani, U. Pagallo, P. Casanovas, and G. Sartor, 40–48. Dordrecht: Springer.

Pagallo, Ugo. 2012b. Robotica. In *Manuale d'informatica giuridica e diritto delle nuove tecnologie*, ed. M. Durante and U. Pagallo, 141–155. Torino: UTET.

Pagallo, Ugo. 2013. What robots want: Autonomous machines, codes, and new frontiers of legal responsibility. In *Human law and computer law: Comparative perspectives*, ed. M. Hildebrandt and J. Gaakeer. Dordrecht: Springer.

Plato. 2006. *The Republic*. Trans. R.E. Allen. New Haven: Yale University Press.

Popper, Karl R. 1935/2002. *The logic of scientific discovery*. London: Routledge.

Popper, Karl R. 1945. *The open society and its enemies*, 2 vols. London: Routledge.

Posner, Richard. 1973. *Economic analysis of law*. Boston: Little Brown (7th ed. 2007 Wolters Kluwer for Aspen Publishers).

Posner, Richard. 1988. The jurisprudence of skepticism. *Michigan Law Review* 86(5): 827–891.

Potter, Norman. 2002. *What is a designer*. London: Hyphen Press.

Rapp, Geoffrey. 2009. Unmanned aerial exposure: Civil liability concerns arising from domestic law enforcement employment of unmanned aerial systems. *North Dakota Law Review* 85: 623–648.

Rasmusen, Eric. 2004. Agency law and contract formation. *American Law and Economics Review* 6(2): 369–409.

Reynolds, Carson, and Masathosi Ishikawa. 2007. Robotic thugs. In *2007 Ethicomp proceedings*, 487–492. Tokyo: Global e-SCM Research Center and Meiji University.

Rezza, Giovanni. 2006. The principle of precaution-based prevention: A Popperian paradox? *European Journal of Public Health* 16(6): 576–577.

Rosenberg, Jeffrey. 2002. Spiders and crawlers and bots, Oh My: The economic efficiency and public policy of online contracts that restrict data collection. *Stanford Technology Law Review* 3, August 19.

Sartor, Giovanni. 2009. Cognitive automata and the law: Electronic contracting and the intentionality of software agents. *Artificial Intelligence and Law* 17(4): 253–290.

Savigny, Frederich. 1979. In *System of the modern roman law*, ed. W. Holloway. Westport: Hyperion.

Scott, Samuel P. (ed.). 1932. *The civil law*. Cincinnati: Central Trust.

Sharkey, Noel. 2008. Grounds for discrimination: Autonomous robot weapons. *RUSI Defence Systems* 11(2): 86–89.

Sharkey, Noel. 2011. Automated warfare: Lessons learned from the Drones. *Journal of Law, Information and Science* 21(2). doi:10.5778/JLIS.2011.21.Sharkey.1.

Sharkey, Noel, Marc Goodman, and Nick Ross. 2010. The coming robot crime wave. *IEEE Computer Society* 43: 114–116.

Shneiderman, Ben. 2000. Universal usability. *Communications of the ACM* 43(5): 84–91.

Singer, Peter. 2009. *Wired for war: The robotics revolution and conflict in the 21st century*. London: Penguin.

Singer, Peter. 2011. A world of killer apps. *Nature* 477: 400.

Smith, Vernon L. 1962. An experimental study of competitive market behaviour. *Journal of Political Economy* 70(2): 111–137.

Solum, Lawrence B. 1992. Legal personhood for artificial intelligence. *North Carolina Law Review* 70: 1231–1287.

Sparrow, Robert. 2007. Killer robots. *Journal of Applied Philosophy* 24(1): 62–77.

Ŝtaerman, Elena M., and Mariana K. Trofimova. 1975. *La schiavitù nell'Italia imperiale. I-III secolo*. Roma: Editori Riuniti.

Sullins, John P. 2011. Introduction: Open questions in roboethics. *Philosophy and Technology* 24(3): 233–238.

Sunder, Shyam. 2004. Markets as artifacts: Aggregate efficiency from zero-intelligence traders. In *Models of a man: Essays in memory of Herbert A. Simon*, ed. M. Augier and J. Marsch, 501–519. Cambridge, MA: MIT Press.

Teubner, Günther. 2007. *Rights of non-humans? Electronic agents and animals as new actors in politics and law*. Max Weber Lecture at the European University Institute of Fiesole, Italy, January 17.

Thorburn, William M. 1917. What is a person? *Mind* 26(103): 291–316.

UN World Robotics. 2005. *Statistics, market analysis, forecasts, case studies and profitability of robot investment*, ed. UN Economic Commission for Europe and co-authored by the International Federation of Robotics, UN Publication, Geneva, Switzerland.

Veruggio, Gianmarco .2006. Euron roboethics roadmap. In *Proceedings Euron Roboethics Atelier*, 27 February–3 March, Genoa, Italy.

Wallach, Wendell, and Colin Allen. 2009. *Moral machines: Teaching robots right from wrong*. New York: Oxford University Press.

Watson, Alan (ed.). 1988. *The digest of Justinian*, vol. I. Philadelphia: University of Pennsylvania Press.

Weber, Max. 1904/1949. Objectivity in social science and social policy. In *The methodology of the social sciences*, eds. and trans. E.A. Shils and H.A. Finch . New York: Free Press.

Wein, Leon E. 1992. The responsibility of intelligent artefacts: Toward an automation jurisprudence. *Harvard Journal of Law & Technology* 6: 103–154.

Weitzenboeck, Emily Mary. 2001. Electronic agents and the formation of contracts. *International Journal of Law and Information Technology* 9(3): 204–234.

Wellman, Michael, Amy Greenwald, and Peter Stone. 2007. *Autonomous bidding agents: Strategies and lessons from the trading agent competition*. Cambridge, MA: MIT Press.

Wiener, Norbert. 1950. *The human use of human beings: Cybernetics and society*. New York: Doubleday.

Wooldridge, Michael J., and Nicholas R. Jennings. 1995. Agent theories, architectures, and languages: A survey. In *Intelligent agents*, ed. M. Wooldridge and N.R. Jennings, 1–22. Berlin: Springer.

WP29. 2009.*The future of privacy*. EU Working Party art.29 D-95/46/EC: WP 168, December 1.

Wu, Stephen S. 2012. Unmanned vehicles and US product liability law. *Journal of Law, Information and Science* 21(2). doi:10.5778/JLIS.2011.21.Wu.1.

Yeung, Karen. 2007. Towards an understanding of regulation by design. In *Regulating technologies: Legal futures, regulatory frames and technological fixes*, ed. R. Brownsword and K. Yeung, 79–108. London: Hart.

Zittrain, Jonathan. 2007. Perfect enforcement on tomorrow's internet. In *Regulating technologies: Legal futures, regulatory frames and technological fixes*, ed. R. Brownsword and K. Yeung, 125–156. London: Hart.

FSC
www.fsc.org
MIX
Papier | Fördert
gute Waldnutzung
FSC® C083411

Kolibri 360 GmbH
Ferdinand-Jühlke-Straße 7
99095 Erfurt, Deutschland
produktsicherheit@kolibri360.de